Hyperspectral Imaging - A Perspective on Recent Advances and Applications

Edited by Jung Y. Huang

Published in London, United Kingdom

Hyperspectral Imaging - A Perspective on Recent Advances and Applications
http://dx.doi.org/10.5772/intechopen.102315
Edited by Jung Y. Huang

Contributors
Paul Steadman, Raymond Fan, K. Priya, K.K. Rajkumar, Samuel Ortega, Karsten Heia, Stein-Kato Lindberg, Kathryn E. Anderssen, Xian-Hua Han, Zhe Liu, Jeanette Hariharan, Yiannis Ampatzidis, Jaafar Abdulridha, Ozgur Batuman

First published in London, United Kingdom, 2023 by IntechOpen
IntechOpen is the global imprint of INTECHOPEN LIMITED, registered in England and Wales, registration number: 11086078, 5 Princes Gate Court, London, SW7 2QJ, United Kingdom

British Library Cataloguing-in-Publication Data
A catalogue record for this book is available from the British Library

Additional hard and PDF copies can be obtained from orders@intechopen.com

Hyperspectral Imaging - A Perspective on Recent Advances and Applications
Edited by Jung Y. Huang
p. cm.
Print ISBN 978-1-83969-093-8
Online ISBN 978-1-83969-094-5
eBook (PDF) ISBN 978-1-83969-113-3

Meet the editor

Jung Y. Huang, a university educator and researcher, has been working to unravel the structures and functional properties of materials and cellular events in living cells with varying optical methodologies. He has co-authored hundreds of journal papers and five book chapters. He also holds tens of patents in laser techniques and single-molecule/hyperspectral imaging and has developed architectural photonics based on hierarchically structured materials. Currently, his research focuses on the integration of artificial intelligence methodology with optics to automatically discover meaningful information from optical sensing/imaging data cubes. He has been an editorial board member and reviewer for several scientific journals. As a member of the global scientific community, he sincerely supports and endeavors to promote the spread of scientific knowledge.

Contents

Preface

Hyperspectral imaging (HSI) is an aerial imaging technology that measures the way an object reflects and emits light at different wavelengths. Typically, it can cover hundreds of bands of light in the electromagnetic spectrum, revealing the precise spectral properties of materials found in the region of interest. With the resulting data, the methodology can distinguish the subtle differences between similar objects, allowing it to map out and differentiate objects and materials in great detail. Due to its fine-grained resolution and ability to distinguish different chemical species, HSI is becoming a powerful tool to spatially resolve the chemistry of materials in varying scientific and engineering disciplines.

HSI data acquisition involves the use of an aerial detector multiplexed in two dimensions and, therefore, requires multiple measurements to complete one data acquisition cycle. The multiple measurements can be executed in two different ways, position scanning or wavelength scanning. Position scanning HSI acquires 2D data of one spatial dimension and the spectral dimension, and scans across the other spatial dimension, whereas wavelength scanning HSI multiplexes the two spatial dimensions and scans across the spectral dimension. Clearly, both methods need time to complete a data cube acquisition. Enabling fast HSI will open doors to new applications where multiple constituents or spatiotemporal dynamics need to be resolved. A variety of snapshot techniques have been developed by invoking a spatial-spectral modulation scheme, such as illuminating an object with a coded light pattern or inserting a spectral modulation module in the HSI imaging device.

HSI is currently applied in many fields. However, we also face a new challenge in data processing and in how to reliably retrieve meaningful information from the high-dimensional HSI data cubes in real-time. Recent progress in both machine-learning and deep-learning techniques may offer a solution to this issue. This book brings together a collection of five chapters offering a glimpse of the status of machine- and deep-learning methodological development for hyperspectral imaging applications.

Chapter 1 "Perspective Chapter: Hyperspectral Imaging for the Analysis of Seafood", by Samuel Ortega et al., presents a survey of current uses of hyperspectral technology for seafood evaluation. The authors briefly describe the optical properties of tissue and offer an introduction to the instrumentation and the developmental status of HSI in the relevant aspects of the seafood industry.

As noted above, consistent data preprocessing and reliable feature extraction are the first step to meaningful information retrieval from high-dimensional data cubes. Chapter 2, "Useful Feature Extraction and Machine Learning Techniques for Identifying Unique Pattern Signatures Present in Hyperspectral Image Data", by Jeanette Hariharan et al., presents a data preprocessing protocol for HSI data. The authors review feature extraction techniques that are useful for identifying pattern signatures embedded in hyperspectral data, and discuss the best practices for processing and analyzing hyperspectral data using machine-learning techniques.

Accurate recording of HSI data ideally requires the data to be acquired with high spatial and spectral resolution. Although until now this has been a fairly lengthy process, deep-learning techniques have recently been developed to improve it. Chapter 3, "Unsupervised Deep Hyperspectral Image Super-Resolution", by Zhe Liu and Xian-Hua Han reviews recent advances in deep unsupervised frameworks for generating high-resolution (HR) HSI, demonstrating with a universally learnable module that only uses low-quality observations to reconstruct the underlying HR-HSI. K. Priya and K.K. Rajkumar, the authors of Chapter 4, "Hyperspectral and Multispectral Image Fusion Using Deep Convolutional Neural Network - ResNet Fusion", point out that in a convolutional neural network (CNN), each layer takes the output from the previous layer, and tends to lose information as the network goes deeper into the architecture. They implement a fusion process in a Residual Network (ResNet) by adding the skip connection between the convolution layers. This skip connection helps to extract more detailed features from the images without any information degradation. The authors measured the results of their ResNet fusion method and found that it exhibits outstanding performance compared with all traditional methods.

In Chapter 5, "Magnetic Scattering with Polarised Soft X-rays", Paul Steadman and Raymond Fan offer an indication of the direction of future HSI development with a proposal for using X-rays as a powerful technique to characterize magnetic materials. Using diffraction, small-angle scattering and reflectivity, the authors demonstrate the element sensitivity and strong dependence of the X-ray polarization on both the size and direction of the magnetic moments theoretically and experimentally.

The use of HSI goes beyond electromagnetic waves, with other available excitation sources such as X-rays and electrons, and new HSI modalities can also be extended to nanometer scales in spatial dimensions. This book brings together diverse HSI research areas to provide a comprehensive overview of the current status of machine- and deep-learning development for hyperspectral imaging.

Jung Y. Huang
Department of Photonics,
Chiao Tung University,
Hsinchu, Taiwan, Republic of China

Section 1

Overview of Hyperspectral Imaging

Chapter 1

Perspective Chapter: Hyperspectral Imaging for the Analysis of Seafood

Samuel Ortega, Stein-Kato Lindberg, Kathryn E. Anderssen and Karsten Heia

Abstract

Hyperspectral imaging technology is able to provide useful information about the interaction between electromagnetic radiation and matter. This information makes possible chemical characterization of materials in a non-invasive manner. For this reason, the technology has been of great interest for the food industry in recent decades. In this book chapter, we provide a survey of the current status of the use of hyperspectral technology for seafood evaluation. First, we provide a brief description of the optical properties of tissue and an introduction to the instrumentation used to capture these images. Then, we survey the main applications of hyperspectral imaging in the seafood industry, including the quantification of different chemical components, the estimation of freshness, the quality assessment of seafood products, and the detection of nematodes, among others. Finally, we provide a discussion about the current state of the art and the upcoming challenges for the application of this technology in the seafood industry.

Keywords: hyperspectral imaging, food quality, seafood industry, spectroscopy, fish

1. Introduction

Hyperspectral imaging is a technology able to measure simultaneously both the spectral and the spatial features of objects or materials under examination. The spectral properties are produced by the interaction between the electromagnetic radiation and the different constituents in a sample, which produces distinct absorption, reflection, and scattering effects on the incident light [1]. The aforementioned optical properties of the different materials are related to their chemical composition and physical properties. Hyperspectral technology for food quality inspection has two main advantages. First is its non-invasive nature, which makes it possible to perform a chemical analysis of the samples without the need to handle them in any way. Secondly, the measurement is very quick to perform as data can be obtained for an entire sample in the matter of seconds. These aspects make the technology easy to integrate with a conveyor belt, which makes it possible to analyze every sample individually. This is preferable to random screening, where the properties of a small batch of sub-samples are analyzed, and it is assumed that their chemical properties are the same for the whole population.

For these reasons, in recent years, hyperspectral imaging has awakened the interest of many researchers for the analysis of food products. According to Scopus, the total number of scientific articles related to studies on hyperspectral imaging for food applications is 1305 in the past 22 years (from 2000 to 2022), with an increasing trend in the number of publications (**Figure 1**).

The range of applications within the food industry is wide and has been extensively covered in the literature by several literature reviews. Those studies cover a wide range of applications including wheat-based products [2], dairy products [3, 4], cereals [5], fruits and vegetables [6, 7], meat [8–10], or condiments [11]. Additionally, it has also been applied to detect adulteration [12] or fraud [13].

Furthermore, other researchers have analyzed the potential of hyperspectral imaging for food microbiology inspection [14] or for the optimization of agricultural procedures [15].

The common motivation for all of these research efforts is to find new technologies able to determine quality parameters on food products, with the goal of avoiding the use of traditional characterization techniques, which are usually destructive, time consuming, and, in certain cases, subjective.

In this book chapter, we provide a survey of the current status on the use of hyperspectral imaging technology in the seafood industry as well as potential future applications. It is worth noting that the workflow for the investigation of hyperspectral imaging in this field requires an appropriate experimental design, the use of adequate instrumentation to carry out data acquisition campaigns, the collection of reference data, and finally the image processing of the hyperspectral images. For this reason, the research performed in this field usually requires a close multidisciplinary collaboration of skilled professionals from different fields, such as biologists, physicists, and engineers, among others.

This book chapter is organized as follows. First, a brief description of the optical properties of different tissue constituents is provided. Second, we discuss the most relevant factors about the instrumentation that should be considered for food inspection applications. Then, we provide a survey about the specific proposed solutions for

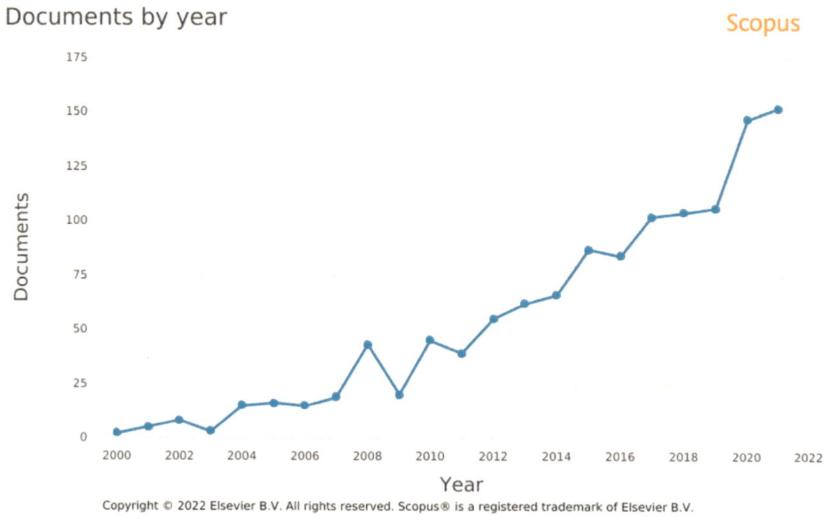

Figure 1.
Scientific publications related to the use of hyperspectral imaging for food applications.

the use of hyperspectral imaging evaluation of seafood products. This survey is not technical, and it has been focused on the goal of providing a description of the wide range of applications that have been covered in the literature until now. We also provide the readers with a summary table containing more specific details of the different research works presented in this book chapter. Finally, we discuss the current limitations of the technology and the potential future trends for hyperspectral imaging use in the seafood industry.

2. Optical properties of biological tissue

The quantification of the chemical constituents of biological tissue is possible due to the optical properties of light when propagating within it. The three types of interactions between electromagnetic radiation and tissue that can be measured are absorption, refraction, and scattering [16]. Light absorption is related to the amount of electromagnetic radiation that is transformed into energy by tissue molecules. The different molecules will present specific absorption peaks, which are related to the transitions between two energy levels by light at specific wavelengths.

The absorption peaks of different biological tissue constituents in the visible and near-infrared regions of the electromagnetic spectrum have been widely characterized in the literature. For that reason, the absorption spectra of water, lipids, proteins, collagen, and hemoglobin in its different oxygenation states are known [17, 18]. A representation of those absorption peaks in the spectral range from 500 to 1600 nm is presented in **Figure 2** [19].

3. Instrumentation

Every hyperspectral acquisition system is composed of a lens, an optical element employed to perform the spectral sampling, an electronic sensor, and a light source. There are different types of hyperspectral systems depending on how the sampling of

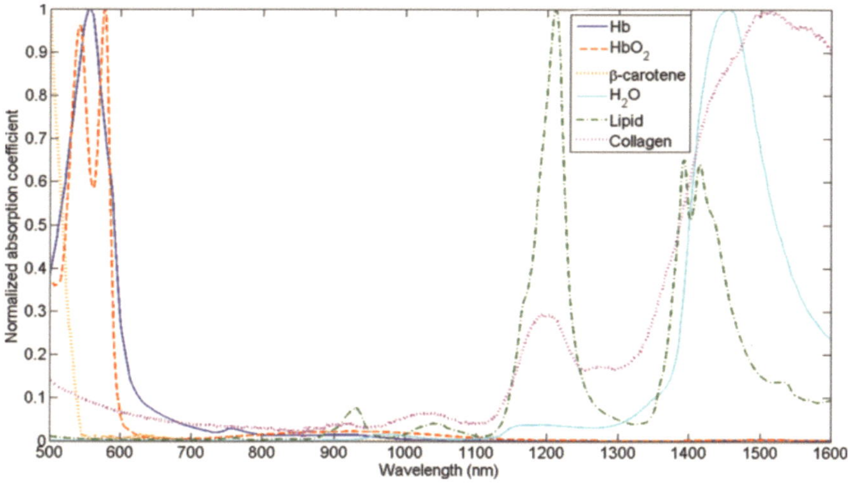

Figure 2.
Absorption peaks of different tissue constituents in the spectral range from 500 to 1600 nm. Reproduced from [19]; creative commons BY 4.0; published by SPIE (2011).

the electromagnetic spectra is accomplished. This information is beyond the scope of this book chapter, but readers who are interested in this can refer to different reviews on hyperspectral imaging hardware in the literature [20].

A relevant characteristic of hyperspectral imaging instrumentation that is extensively mentioned in this book chapter is the spectral range. The spectral range defines the region of the electromagnetic spectrum that a hyperspectral camera is able to measure. In commercial hyperspectral systems, there are standard definitions for the spectral range. Visible and near-infrared (VNIR) refers to the spectral range from 400 to 1000 nm, while near-infrared (NIR) and short-wave infrared (SWIR) are used for the ranges 1000–1700 nm and 1000–2500 nm, respectively. Other key parameters in hyperspectral imaging instrumentation are the spectral resolution and the spatial resolution, but these concepts will not be used in this book chapter.

Although the details of hyperspectral cameras are not relevant for this book chapter, the selection of the illumination type to produce the appropriate light–tissue interactions within the sample is relevant. Using a diffuse reflectance illumination scheme, the light is evenly delivered to the sample, and it is measured by a hyperspectral camera after being reflected off its surface. With this illumination mode, the interaction of light and matter is only measured from the surface of the sample. In some cases, the diffuse light can penetrate a small distance into the sample depending on its translucency. However, in complex and inhomogeneous samples, this type of illumination is not enough for accurate characterization of their chemical composition [21, 22]. For this reason, some researchers have proposed the use of interactance (also known as transflectance) illumination, where the light is able to penetrate deeper into the sample. This illumination mode consists of a focused light illuminating the sample in a different spatial location to where the spectral information is captured, allowing the hyperspectral camera to measure the light interaction after multiple internal reflections have occurred inside the sample [23]. In the applications mentioned in this book chapter, both types of light illumination schemes are used.

4. Applications of hyperspectral imaging in the seafood industry

4.1 Chemical composition

The analysis of the chemical composition of seafood products is important for the determination of their overall quality or nutritional value, among others. However, conventional chemical analysis techniques are destructive and time consuming. For that reason, in recent years, hyperspectral imaging has been foreseen as a technology suitable for providing a non-invasive measurement of those chemical properties.

For example, in Atlantic salmon, moisture and fat content are considered to be closely related to the overall quality of the product. The fat content has consequences for both the customers and the industry. For the customers, the amount of fat present in a fresh fillet determines the flavor and texture of the product. For the industry, it is important to quantify the amount of fat in a salmon fillet to determine its target market. For example, the optimum fat content for smoked salmon is between 8 and 12% [24], while salmons with higher fat content and marbling are preferred for sushi and sashimi [25, 26]. Similarly, the moisture is related to the shelf-life of seafood products.

Several research studies have been focused on non-invasive determination of moisture and fat using hyperspectral imaging. Several authors have proposed using NIR spectroscopy to estimate fat and moisture in Atlantic salmon. Zhu *et al.* obtained accurate models using only the spectral information of the samples [27]. However, fat content is not uniform throughout a sample, and Zhang *et al.* demonstrated that more robust models for fat and moisture can be obtained if texture features extracted from characteristic spectral bands are used as predictors [28]. Using the aforementioned approaches, the authors not only predicted the overall fat and moisture content for the samples but also provided their spatial distribution within the salmon fillets (**Figure 3**). In a more technical approach, Dixit *et al.* performed a comparison between two different hyperspectral technologies (line scan and snapshot) working in different spectral ranges for the determination of fat in Atlantic salmon [29]. The authors concluded that the spectral range from 670 to 950 nm was able to provide an equivalent performance in the prediction of fat compared to the spectral range from 550 to 1700 nm, which may lead to the use of cheaper instrumentation for this application due to the narrower spectral range needed.

Another important quality indicator for fish is blood content. During capture, fish are, as a rule, drained of their blood by cutting through the gills. This is mainly done in order to kill the fish quickly, but it also has the effect of preventing the blood from settling in the muscle and changing its color. The appearance of a fish fillet impacts its perceived quality, and a red hue in a whitefish fillet can be off-putting to the consumer. In the case of smoked products, any remaining blood turns brown and can, for instance, be perceived as dark spots in a smoked salmon fillet.

Skjelvareid *et al.* demonstrated that hyperspectral imaging can detect and quantify blood in whitefish fillets [30]. The hemoglobin in the blood absorbs light very strongly in a specific region of the visual spectrum and therefore stands out against the white fish muscle. The different oxidation states of the hemoglobin can also be distinguished

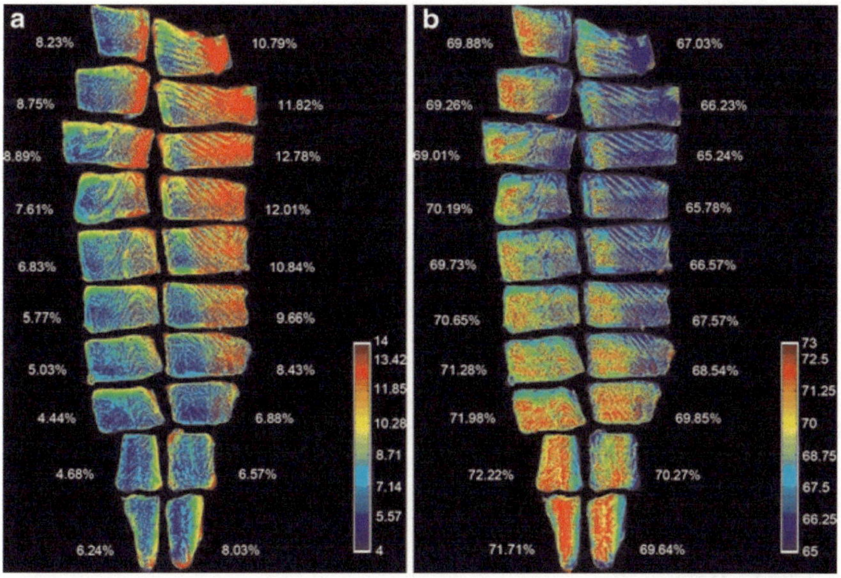

Figure 3.
Spatial mapping of moisture (a) and fat (b). Reprinted by permission from springer nature customer service Centre GmbH: Springer nature, food and bioprocess technology [27] [COPYRIGHT: Springer nature] (2013).

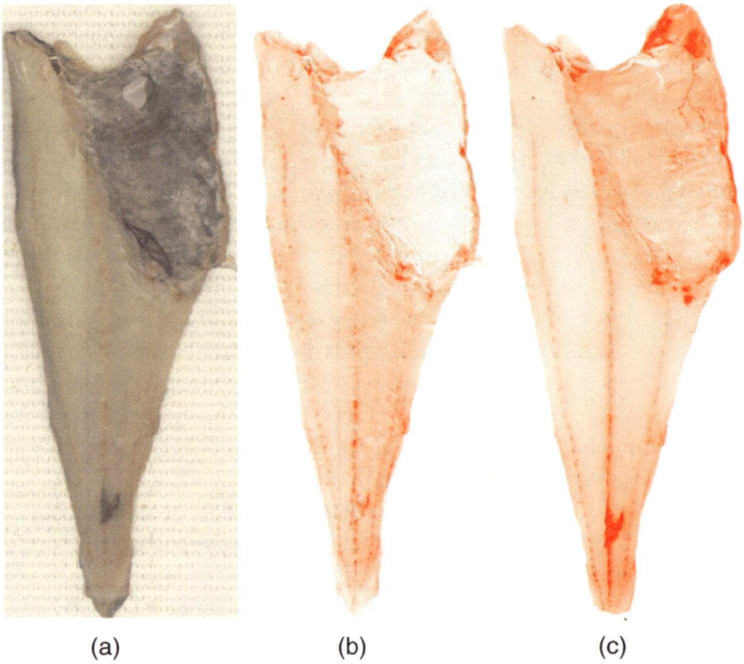

Figure 4.
Quantification of blood in cod fillets using hyperspectral imaging. a) Calibrated color image based on diffuse reflectance hyperspectral imaging. b) Estimated blood concentration based on diffuse reflectance hyperspectral imaging. c) Estimated blood concentration based on interactance hyperspectral imaging [31].

by their spectral signature, which makes it possible to do a pixelwise spectral unmixing by using the known reference spectra for the hemoglobin. An example of the quantification of blood in Atlantic cod fillets can be observed in **Figure 4**.

The same method has been applied to salmon fillets as well. The pigments in the salmon muscle absorb light in the same spectral region as the hemoglobin but with a different spectral profile. It is therefore possible to distinguish the blood from the pigments by taking both of them into account.

Two illumination setups are presented in the above publications. The first one is a diffuse illumination for reflectance imaging, while the other is an interactance. The idea is that surface reflection does not give enough information about the internal properties of the fillet, such that the focused light source of the interactance setup is necessary to penetrate further into the muscle. To ensure that the light recorded by the camera has propagated through the muscle and been attenuated by it, the focused light source is placed a certain distance from the field of view of the camera, which reduces surface reflection in the camera field of view while providing a good signal from the inside of the fillet [31]. This technique has also been shown to work for quantifying blood in whole whitefish through the skin, which, at the time of writing, is being developed into a commercial quality control method [32].

4.2 Analysis of freshness

Technologies able to non-invasively estimate the freshness of seafood products are in demand for the industry. There are currently different techniques for the estimation of freshness in seafood products; however, such methods are labor intensive and usually destructive and cannot be applied to every specimen in the product line. The possibility of technology able to perform rapid freshness analysis for every sample could bring to the industry new alternatives for decision making with the goal to improve the processing and sorting of the raw materials.

Several researchers have investigated the estimation of the freshness of seafood products using hyperspectral imaging. Usually, the approaches followed by those researchers consist of the utilization of spectral data together with multivariate analysis methods to predict the values of different reference measurements related to the freshness.

A basic common reference method for the estimation of freshness is the storage time. Some researchers have successfully estimated storage time as a freshness indicator for fillets from different fish species using hyperspectral imaging, for example, pearl gentian grouper [33], Atlantic salmon [34], and Atlantic cod [35]. Kimiya *et al.* [34] and Sivertsen *et al.* [35] attributed the spectral changes between the different storage times to the oxidation of hemoglobin and myoglobin proteins during the chilled storage, which enables the successful estimation of the storage time based on the spectral information.

The total volatile basic nitrogen (TVB-N) is often used as a biomarker of protein and amine degradation and is considered a proxy freshness of fresh meat and fish products [36]. TVB-N has been widely used as a reference value for freshness estimation using hyperspectral imaging. In the literature, TVB-N estimation in fillets from different species can be found, including rainbow trout [37], grass carps [38, 39], or tilapia [40]. **Figure 5** shows the spatial distribution of TVB-N values within grasp carp fillets. All the above research presented accurate models for predicting TVB-N values using the VNIR spectral range. However, Yu *et al.* demonstrated that combining the VNIR and NIR spectral ranges resulted in improved estimations [40].

Although storage time and TVB-N methods have been the more common reference methods for determining freshness using hyperspectral imaging, other researchers have used alternative methods with successful results. Zhang *et al.* used electrical conductivity on largemouth bass fillets [41], while sensory evaluation of the shelf-life was used as a reference method for the estimation of freshness by Khoshnoudi-Nia *et al.* [42].

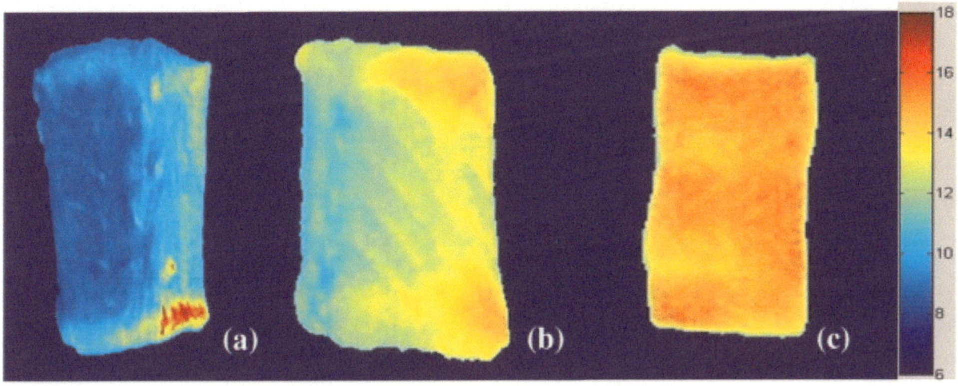

Figure 5.
Spatial distribution of TVB-N values for freshness estimation. (a), (b) and (c) shows the TVB-N for different fillets (8.26, 12.98, and 15.69 mg N/100 g, respectively). Reprinted from innovative Food Science & Emerging Technologies, 21, Jun-Hu Cheng, Da-Wen Sun, Xin-An Zeng, Hong-Bin Pu, non-destructive and rapid determination of TVB-N content for freshness evaluation of grass carp (Ctenopharyngodon idella) by hyperspectral imaging [38], page 9, Copyright (2014), with permission from Elsevier.

4.3 Quality characterization

Quality evaluation of seafood products, and food products in general, is mainly determined by how the odor, color, and texture of the product is perceived by customers. Traditionally, this quality evaluation has been addressed by sensory evaluation panels, who are a group of people trained to perform a quality judgment of seafood products. In recent years, some solutions based on hyperspectral imaging have been investigated to produce objective measurements of these quality parameters for different seafood products to help the industry stakeholders optimize their production.

Texture is a significant feature for the quality perception of seafood products by customers. For the texture evaluation, there are instruments that allow one to perform objective measurements, which are more repeatable than the subjective opinion of a sensory panel. However, the use of texture analyzers is time consuming and destructive. For this reason, some researchers have proposed the use of hyperspectral imaging for the characterization of texture features in seafood products. In those studies, the reference texture data are usually collected using a variety of mechanical instruments able to measure the force needed to compress or tear a sample. Wang *et al.* developed multivariate regression models based on the spectral data from commercial crisp grass carp (*Ctenopharyngodon idellus*) fillets to predict their hardness attributes using the spectral information in the VNIR spectral range [43]. Another research study demonstrated that the use of hyperspectral images in the SWIR spectral range is also suitable for the estimation of texture features in rainbow trout (*Oncorhynchus mykiss*) fillets [44]. The results of these studies showed high correlation between the predicted texture values from the spectral data and the texture measurements. In another innovative study, the authors also obtained promising models for the estimation of texture parameters of fish by using spectral and textural data from eyes and gills [45]. This approach has the advantage of being able to predict the texture of the fish before it is cut into fillets.

Wang *et al.* proposed the use of artificial neural networks together with VNIR spectral data for the characterization of color in large yellow croaker (*Larimichthys crocea*) fillets [46]. In this study, the color variations in the samples were produced by storing the samples in different conditions and acquiring hyperspectral data. The corresponding reference measurements used a colorimeter to quantify the color parameters of the sample. The results of this study showed that hyperspectral imaging is a potential tool for color characterization of samples, with some advantages over the colorimeters. Colorimeters require point measurements, which present two main disadvantages: there is a need for physical contact with the sample, and the measurements are performed in a limited number of spots on the sample.

4.4 Detection of nematodes

Parasites in fish are a significant problem for seafood producers and consumers, presenting both quality and health concerns. Typically, the presence of parasites in products leads to rejection of the product by both purchasers and sellers. Parasites, such as *Anisakis simplex* and *Pseudoterranova decipiens,* are commonly present in whitefish fillets [47]. Today, every single fillet is inspected by transillumination on candling tables [48], and nematodes are removed manually. The detection rate using candling tables has been reported as low as 23% in a recent study by Mercken *et al.* [49]. Manual screening for parasites is an expensive operation previously reported to account for half of the production cost for Pacific cod from the Bering Sea and the Gulf of Alaska [50]. Several different instrumental methods have been evaluated for nematode detection:

fluorescence [51], ultrasonic waves [52], X-ray and computer tomography [31], and multispectral imaging [53]. The first conceptualization on the use of spectroscopic techniques for nematode detection was proposed by Pau *et al.* in 1991, where the spectral differences between the parasites and the fish muscle were shown [54]. The chemical differences between nematodes and fish muscle were documented by Stormo *et al.* [55], and a later work discussed the impact of selecting a limited number of wavelengths based on such chemical differences [56]. In Sigernes *et al.*, the authors developed a custom spectral imager targeting a wide variety of seafood industry applications [57]. In that work, the authors showed as a proof of concept that the spectral information can be potentially used to identify nematodes in fish samples. Using the same instrumentation, Heia *et al.* conducted the first research study on the detection of nematodes with hyperspectral images [58]. Using the transmittance illumination mode, this work served as a proof of concept to show the potential of spectral imaging for nematode detection. The goal of using transillumination was to be able to detect nematodes deeply embedded in the fish flesh. However, this preliminary work was limited by a low number of samples and ideal laboratory conditions. With the goal of making the system more suitable for an industrial setting, Sivertsen *et al.* further investigated this research line. First, a transillumination setup based on a commercial hyperspectral camera with a higher number of samples was evaluated [59]. However, despite the promising results in the detection of nematodes, the transmittance setup still presented obstacles for implementation in industry, for example, a low imaging speed and challenges regarding the optimization of the light conditions. For those reasons, in a subsequent study, Sivertsen *et al.* proposed for the first time the use of interactance hyperspectral imaging for the detection of nematodes [60]. In this work, the authors were able to satisfy industrial needs for fast acquisition and processing of the images. However, although the detection rate of nematodes was comparable with the human manual inspection, the detection rate was still low and the false positive rate too high to meet industrial requirements.

In an ongoing research project funded by the Norwegian Seafood Research Fund (FHF), entitled *Commercial Nematode Detection in Whitefish Fillets* (901614), a solution is being developed to perform nematode detection using hyperspectral imaging. The project is being conducted by the Norwegian Institute of Food, Fisheries and Aquaculture Research (Nofima) and Maritech, a company commercializing a hyperspectral solution for seafood inspection called Maritech Eye™. In the previous approaches, only the spectral information from the nematodes and the fish muscle was exploited. In this project, a solution based on a deep learning neural network, where both the spatial and the spectral features of the data are utilized to detect the nematodes, is proposed. **Figure 6** shows the manual annotation of the nematodes as well as the automatic detection of the nematodes using hyperspectral image analysis with a deep neural network. The experiment to demonstrate the feasibility of this approach was tested under industrial conditions in a cod production factory belonging to the company Maredeus (Portugal). The results of the proposed approach were accurate, with a high detection rate and almost no false positives. In addition, the system was able to operate at industrial speed (400 mm/second), including both the image acquisition and the data analysis, which would make it possible to use this approach as an industrial solution for the detection of nematodes.

4.5 Identification of different species

Another challenge for seafood production lines is the automatic sorting of different species when they are processed simultaneously. Additionally, the use of imaging

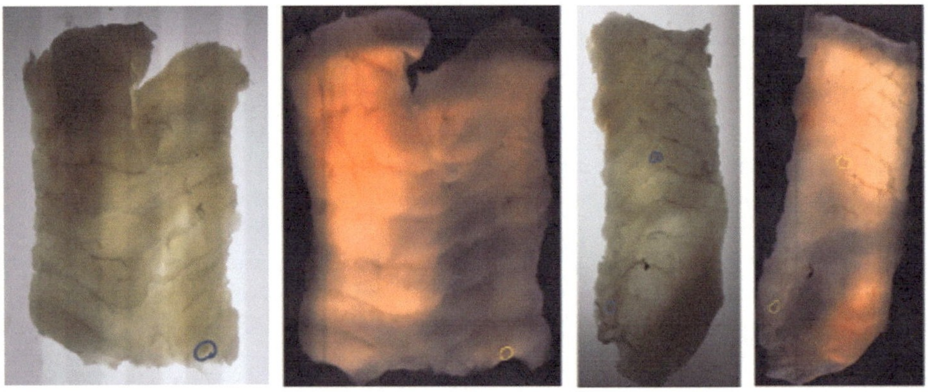

Figure 6.
Manual annotation of nematodes (blue) and automatic prediction (yellow) of their location on cod fillets based on hyperspectral image analysis.

technologies able to identify different fish species is attractive both for the consumer and for the industry, since they can help to mitigate fraud in fish mislabeling [61]. In a research study performed by Chauvin *et al.*, the authors evaluated the potential of the spectral information of fillets from different species in order to correctly classify them [62, 63]. A total of 22 fish species were recorded using diffuse reflectance illumination (VNIR and SWIR spectral ranges) and fluorescence excitation (VIS). Using this data, different supervised classifiers based on the spectral data from the different species were trained. The results obtained in this study suggest that the combination of spectral channels from the different spectral ranges and imaging modalities improve the classification compared to single-mode data (i.e., only VNIR, only SWIR, or only fluorescence). Finally, the authors investigated reducing the number of spectral bands needed for species identification without compromising on the performance of the classifier. The outcome of this research was a selection of 7 spectral bands that can be potentially used for the identification of species. This finding paves the way for the future development of cheap instruments based on LED illumination using such specific wavelengths to perform the species sorting.

Beyond the seafood industry, hyperspectral imaging has also been investigated for species identification with the goal of using this technology as a complementary tool to existing molecular and morphological techniques for faunal biodiversity assessment. Kolmann *et al.* performed a study in South American fish species that are difficult to distinguish even under controlled conditions: piranhas and pacus (both from the family Serrasalmidae) [64]. The authors were able to successfully discriminate between 47 different species and subspecies, using only their spectral information (**Figure** 7). The outcomes of this study demonstrated hyperspectral imaging as a potential technology for biodiversity screening.

4.6 Damage detection

One of the main pretreatments applied to freshwater fish is scale removal; however, methods to do this can produce damage to the product. With the goal of better characterizing the damages caused by the different physical scale removal methods, Wang *et al.* proposed the utilization of VNIR spectral data as a tool to visualize such

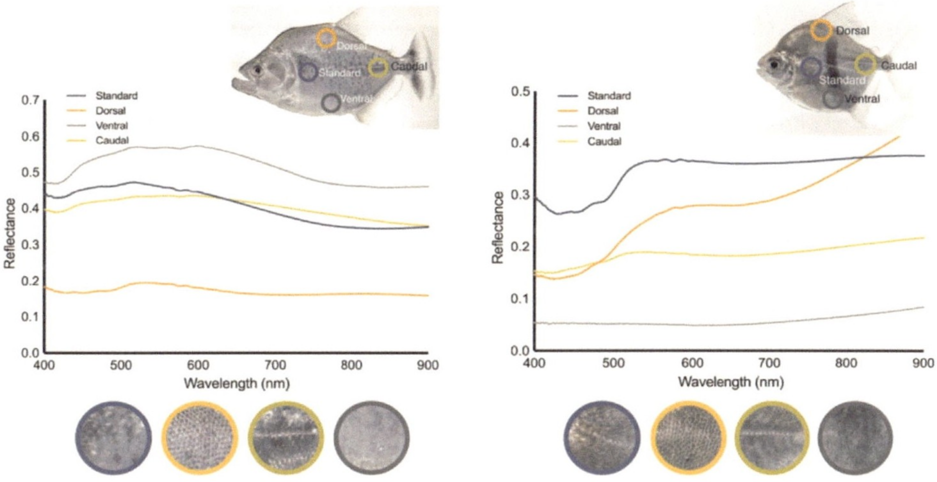

Figure 7.
*Comparative spectral signatures from four different body regions of a pacu (*Myloplus schomburgkii*) (right) and a piranha (Serrasalmus geryi) (left). Reproduced from [64]; creative commons BY 4.0; published by springer nature limited (2021).*

damages [65]. The results of this study were positive, showing an accurate identification of the damaged areas based only on the spectral information.

Another type of damage occurs when fish are caught. Jensen *et al.* proposed the use of a catch damage index based on VNIR hyperspectral information to characterize the catch damage when different trawling strategies are used [66]. The method is based on the estimation of the residual blood in fish muscle by using constrained spectral unmixing [30]. Using this application of hyperspectral image processing, it was possible to conduct an experiment to evaluate the effect of different trawling strategies on fish damage.

4.7 Detection of contamination agents

Plastic contamination in marine environments leads to the ingestion of microplastics by fish. There is evidence that indicates that microplastics intake causes harmful effects to fish health [67]. In recent years, research has shown an increasing trend in the presence of microplastics in seafood products [68]. However, the methods to accurately quantify the presence of microplastics are complex and expensive, which complicates the experimental trials required to quantify the effect of this problem. For this reason, Zhang *et al.* proposed the use of hyperspectral imaging in the range from 900 to 1700 nm for the identification of microplastics [69]. With the goal of training a supervised classifier based on the spectral information, the intestinal tract contents of different fish were contaminated with plastic polymers of different chemical composition, size, and color. Then, the accuracy of the proposed methodology was evaluated, both in prepared samples and in fish samples from three different species. The results of this experiment indicated that hyperspectral imaging can be a suitable technology to detect the presence of microplastics in the intestinal tract of fish. However, the precision in the detection is affected by the size of the plastic particles, which makes it necessary to increase the dataset to improve the machine learning models to improve the detection of small plastic particles.

From the food safety perspective, the detection of harmful microorganisms present in fish is a relevant topic. With the goal of developing imaging technologies for the detection of *Enterobacteriaceae* contamination in Atlantic salmon (Salmo salar) flesh, He *et al.* investigated the use of the NIR spectra for monitoring the presence of such bacteria [70]. After capturing hyperspectral images of salmon contaminated with *Enterobacteriaceae* at different storage periods, the authors were able to quantify the presence and severity of the bacterial contamination. It is worth noticing that hyperspectral imaging technology is not able to measure the bacteria presence by itself; however, there are differences between the spectra from contaminated and non-contaminated salmon flesh.

4.8 Applications in aquaculture

Aquaculture production has significantly grown during the past 20 years [71]. This is mainly due to the increasing demand for seafood products, together with the goal of the seafood industry to increase productivity. Thus, there is a current demand for novel information and digital technologies that can be applied in aquaculture to improve the productivity of fish farms [72]. Nowadays, the use of hyperspectral imaging technologies in aquaculture is limited to a few contributions.

In Atlantic salmon (S. salar) farming, the transition from juvenile freshwater fish (parr) to seawater adapted fish (smolt) is called smoltification. Smoltification involves changes in the morphology, physiology, and biochemistry of juvenile salmon. From a fish farmer perspective, it is important to monitor the smoltification process for two principal reasons. On the one hand, an incomplete smoltification process at the time the salmon is transferred to seawater leads to poor salmon welfare and an increased risk of mortality. On the other hand, a late transition to seawater generates negative consequences for the farmer since the production chain is not optimized, which induces economic losses. With the goal of providing the aquaculture industry with a solution to this problem, Svendsen *et al.* studied the relationship between the spectral information and the physiological changes in juvenile salmons [73]. After analyzing more than 300 fish from three different farms, the authors were able to perform an accurate discrimination between parr and smolt with high sensitivity and specificity. The classification was performed using a machine learning classifier (Support Vector Machine) using only three specific spectral channels.

Salmon lice (*Lepeophtheirus salmonis*) are parasites that live on salmonid fishes. The salmon lice represent a huge problem for both farmed and wild salmon because they can produce severe fin damage, bleeding, and open wounds in the host. Salmon affected by these parasites are likely sensitive to other pathogens, leading to increased sickness. Therefore, salmon lice are a problem that generates negative effects in salmon welfare and leads to significant economic losses for the farmers and suffering for the fish. Early warnings for lice could help farmers to take action to eliminate the infestation. However, the identification and counting of these parasites are challenging tasks even for skilled staff. With the goal of providing an automatic solution to this problem, Pettersen *et al.* conducted an experiment where underwater hyperspectral imaging was applied for the detection of salmon lice [74]. First, the authors recorded and characterized the spectral signature of different salmon lice subtypes in laboratory conditions, in both air and underwater conditions. Finally, they tested the method to identify the different lice subtypes in salmon using the underwater hyperspectral imaging system. Although the research suggested underwater hyperspectral imaging as a promising technology for the detection of salmon lice in sea cages, it can be considered as a proof of concept, and more research needs to be performed to optimize the instrumentation for use as a final product in aquaculture farms.

The use of hyperspectral imaging has also been applied in an indirect manner with the goal to improve the quality of the fish feed in aquaculture. Marine fishmeal powder is added as protein supplement in fish feed in aquaculture, but recently, the adulteration of this product with cheaper alternatives with lower nutritional value has become a common trend. To address this fraud, Kong *et al.* proposed the use of NIR hyperspectral imaging and convolutional neural networks for the identification of adulterants in marine fishmeal [75].

The aforementioned examples suggest that hyperspectral imaging technology can contribute to improvements in aquaculture in the near future.

5. Summary table

In this section, we provide a summary of the main research works that have been covered in this book chapter. In **Table 1**, the information about each research work is specified. This information includes the type of application, the fish species, the type of samples, the number of specimens, the illumination modality, the spectral range, and the image-processing method used to retrieve information from the hyperspectral images.

Application	Fish Species	Sample Type	N	Illum.	Spectral Range	Processing Method	Ref.
Chemical composition (fat and moisture)	Atlantic Salmon	Fillets	5	DR	900–1700 nm	PLSR	[27]
Chemical composition (fat and moisture)	Atlantic Salmon	Fillets	10	DR	900–1700 nm	Optimal Band Selection, Texture Feature Extraction, Multivariate Regression Algorithms	[28]
Chemical composition (fat)	Atlantic Salmon	Fillets	45	DR	550–1700 nm 470–630 nm 670–950 nm	PLSR	[29]
Chemical composition (blood)	Atlantic Cod	Homogenized samples	9	IA	430–1000 nm	Constrained Spectral Unmixing	[30]
Freshness estimation (storage time)	Pearl Gentian Grouper	Fillets	22	DR	900–1700 nm	PLSR	[33]
Freshness estimation (storage time)	Atlantic Salmon	Fillets	48	IA	400–1000 nm	PLSR	[34]
Freshness estimation (storage time)	Atlantic Cod	Fillets	49	IA	400–1000 nm	PLSR	[35]

Application	Fish Species	Sample Type	N	Illum.	Spectral Range	Processing Method	Ref.
Freshness estimation (TVB-N)	Grass Carp	Fillets	30	DR	400–1000 nm	PLSR, Optimal Band Selection	[38, 39]
Freshness estimation (TVB-N)	Tilapia	Fillets	40	DR	400–1000 nm 900–1700 nm	Spectral data fusion, Multivariate Regression Algorithms	[40]
Freshness estimation (sensory analysis)	Largemouth Bass	Fillets	20	DR	400–1000 nm	PLSR, Optimal Band Selection	[41]
Freshness estimation (sensory analysis)	Rainbow Trout	Fillets	40	DR	430–1010 nm	Multivariate Regression Algorithms	[42]
Texture characterization	Crisp Grass Carp	Fillets	15	DR	400–1100 nm	Multivariate Regression Algorithms	[43]
Texture characterization	Rainbow Trout	Fillets	80	DR	1000–2500 nm	PLSR	[44]
Texture characterization	Crucian Carp	Fillets and Whole Fish	84	DR	900–1700 nm	PLSR, Spectral and Textural Feature Extraction	[45]
Color characterization	Large Yellow Croaker	Fillets	15	DR	400–1000 nm	Artificial Neural Networks	[46]
Nematode detection	Atlantic Cod	Fillets	8	IA	350–950 nm	PLS-DA	[58]
Nematode detection	Atlantic Cod	Fillets	40	IA	400–1000 nm	Fisher Discriminant Ratio	[59]
Nematode detection	Atlantic Cod	Fillets	43	IA	400–1000 nm	Custom local calibration, Fisher Linear Classifier	[60]
Identification of species	22 Different Species	Fillets	133	DR FL	419–1007 nm 438–718 nm 842–2532 nm	Artificial Neural Networks, Band selection methods	[62]
Identification of species and subspecies	Piranhas and Pacus	Live fish	176	DR	325–1075 nm	Adaptative Coherence Estimator	[64]
Damage characterization (scale removal)	Fresh Carp	Whole fish	50	DR	387–1024 nm	Decision Trees, Self-Organizing Maps	[65]
Damage characterization (catch damage)	Atlantic Cod	Whole Fish	600	IA	400–1000 nm	Constrained Spectral Unmixing	[66]

Application	Fish Species	Sample Type	N	Illum.	Spectral Range	Processing Method	Ref.
Detection of microplastics	Sea Bass Redeye Mullets Goosefishes	Internal organs	20	DR	900–1700 nm	SVM classification	[69]
Detection of bacteria contamination	Atlantic Salmon	Fillets	30	DR	900–1700 nm	PLS, Optimal Band Selection	[70]
Smoltification monitoring	Atlantic Salmon	Live Fish	314	DR	400–1000 nm	SVM classification	[73]
Salmon lice detection	Atlantic Salmon	Live Fish	7	DR	370–800 nm	PLS-DA, SAM	[74]

*N: Number of samples, **Illum:** Illumination Mode (**DR:** Diffuse Reflectance, **IA:** Interactance, **FL:** Fluorescence). **PLSR:** Partial Least Squares Regression, **SVM:** Support Vector Machines, **PLS-DA:** Partial Least Squares Discriminant Analysis, **SAM:** Spectral Angle Mapper.*

Table 1.
Summary table.

6. Conclusions

In this book chapter, we have surveyed the main applications of hyperspectral imaging for seafood industry-related problems. The main goal in most of the research carried out in this field is to provide an alternative to the expensive, time-consuming, and invasive reference methods that are currently employed for the characterization of seafood products. Additionally, the advantage of hyperspectral technology is its applicability to industrial production chains, where the analysis can be performed individually for every sample, which can lead to the optimization of production and decision making for the industry.

Although the application field of this technology is wide and promises to address actual problems for both the industry and the consumers, there are still challenges that must be carefully investigated in the upcoming years.

As far as the instrumentation is concerned, there are still uncertainties about which type of illumination (diffuse reflectance or interactance) is more appropriate for each application. Additionally, there is no strict criterion for the selection of the most adequate spectral range for each application. In this sense, more comparative research should be carried out in order to have clearer arguments on which spectral range should be used for different applications.

The number of processing methods used to extract information from hyperspectral data is huge and diverse. An appropriate evaluation of these methods should be carefully carried out to gain a better understanding of their limitations and advantages for each scenario. Additionally, most of the methods covered in the literature are based exclusively on the spectral information, while the spatial information is usually underrated. However, the trend in hyperspectral image analysis in other fields is to try to exploit simultaneously both the spatial and the spectral features of the data, especially with the rise of sophisticated deep learning architectures to this end [76, 77].

Regarding the future of this research line, the upcoming challenges should be focused on the transfer of knowledge to industry, where this technology could be employed to improve production chains and decision making. In this sense, commercial products consisting of industrial-grade spectral imaging systems have been recently launched, such as the QMonitor (QVision AS, Oslo, Norway) or the Maritech Eye (Maritech Systems AS, Molde, Norway). Both systems are based on interactance illumination mode. The QMonitor is a multispectral NIR system, while the Maritech Eye is a hyperspectral system in the VNIR spectral range. These devices have been proven to be useful for different food quality applications [78–82] and are currently used in food industry production facilities.

Acknowledgements

This project is supported by DigiFoods, a Norwegian Strategic Research Initiative (project number 309259), and is also part of NFR funding grant 294805.

Conflict of interest

The authors declare no conflict of interest.

Author details

Samuel Ortega*, Stein-Kato Lindberg, Kathryn E. Anderssen and Karsten Heia
Nofima, Norwegian Institute of Food Fisheries and Aquaculture Research, Tromsø, Norway

*Address all correspondence to: samuel.ortega@nofima.no

References

[1] Lu R, editor. Light Scattering Technology for Food Property, Quality and Safety Assessment. London: CRC Press Taylor & Francis Group; 2017

[2] Badaró AT, Hebling e Tavares JP, Blasco J, Aleixos-Borrás N, Barbin DF. Near infrared techniques applied to analysis of wheat-based products: Recent advances and future trends. Food Control. 2022;**140**:109115. DOI: 10.1016/j.foodcont.2022.109115

[3] Hebling e Tavares JP, da Silva Medeiros ML, Barbin DF. Near-infrared techniques for fraud detection in dairy products: A review. Journal of Food Science. 2022;**87**(5):1943-1960. DOI: 10.1111/1750-3841.16143

[4] Bittante G, Patel N, Cecchinato A, Berzaghi P. Invited review: A comprehensive review of visible and near-infrared spectroscopy for predicting the chemical composition of cheese. Journal of Dairy Science. 2022; **105**(3):1817-1836. DOI: 10.3168/jds.2021-20640

[5] An D, Zhang L, Liu Z, Liu J, Wei Y. Advances in infrared spectroscopy and hyperspectral imaging combined with artificial intelligence for the detection of cereals quality. Critical Reviews in Food Science and Nutrition. 2022:1-31. DOI: 10.1080/10408398.2022.2066062

[6] He Y, Xiao Q, Bai X, Zhou L, Liu F, Zhang C. Recent progress of nondestructive techniques for fruits damage inspection: A review. Critical Reviews in Food Science and Nutrition. 2022;**62**(20):5476-5494. DOI: 10.1080/10408398.2021.1885342

[7] Lu Y, Saeys W, Kim M, Peng Y, Lu R. Hyperspectral imaging technology for quality and safety evaluation of horticultural products: A review and celebration of the past 20-year progress. Postharvest Biology and Technology. 2020;**170**:111318. DOI: 10.1016/j.postharvbio.2020.111318

[8] Antequera T, Caballero D, Grassi S, Uttaro B, Perez-Palacios T. Evaluation of fresh meat quality by hyperspectral imaging (HSI), nuclear magnetic resonance (NMR) and magnetic resonance imaging (MRI): A review. Meat Science. 2021;**172**:108340. DOI: 10.1016/j.meatsci.2020.108340

[9] Jia W, van Ruth S, Scollan N, Koidis A. Hyperspectral imaging (HSI) for meat quality evaluation across the supply chain: Current and future trends. Current Research in Food Science. 2022; **5**:1017-1027. DOI: 10.1016/j.crfs.2022.05.016

[10] Falkovskaya A, Gowen A. Literature review: Spectral imaging applied to poultry products. Poultry Science. 2020; **99**(7):3709-3722. DOI: 10.1016/j.psj.2020.04.013

[11] Mei J, Zhao F, Xu R, Huang Y. A review on the application of spectroscopy to the condiments detection: From safety to authenticity. Critical Reviews in Food Science and Nutrition. 2022;**62**(23):6374-6389. DOI: 10.1080/10408398.2021.1901257

[12] Faith Ndlovu P, Samukelo Magwaza L, Zeray Tesfay S, Ramaesele Mphahlele R. Destructive and rapid non-invasive methods used to detect adulteration of dried powdered horticultural products: A review. Food Research International. 2022;**157**:111198. DOI: 10.1016/j.foodres.2022.111198

[13] Nobari Moghaddam H, Tamiji Z, Akbari Lakeh M, Khoshayand MR, Haji

Mahmoodi M. Multivariate analysis of food fraud: A review of NIR based instruments in tandem with chemometrics. Journal of Food Composition and Analysis. 2022;**107**: 104343. DOI: 10.1016/j.jfca.2021.104343

[14] Soni A, Dixit Y, Reis MM, Brightwell G. Hyperspectral imaging and machine learning in food microbiology: Developments and challenges in detection of bacterial, fungal, and viral contaminants. Comprehensive Reviews in Food Science and Food Safety. 2022; **21**(4):3717-3745. DOI: 10.1111/1541-4337.12983

[15] Khan A, Vibhute AD, Mali S, Patil CH. A systematic review on hyperspectral imaging technology with a machine and deep learning methodology for agricultural applications. Ecological Informatics. 2022;**69**:101678. DOI: 10.1016/j.ecoinf.2022.101678

[16] Tuchin VV. Tissue Optics: Light Scattering Methods and Instruments for Medical Diagnosis. 3rd ed. Bellingham WA: SPIE; 2015

[17] Wilson RH, Nadeau KP, Jaworski FB, Tromberg BJ, Durkin AJ. Review of short-wave infrared spectroscopy and imaging methods for biological tissue characterization. Journal of Biomedical Optics. 2015;**20**(3):030901. DOI: 10.1117/1.JBO.20.3.030901

[18] Khodabux K, Lomelette M, Jhaumeerlaulloo S, Ramasami P, Rondeau P. Chemical and near-infrared determination of moisture, fat and protein in tuna fishes. Food Chemistry. 2007;**102**(3):669-675. DOI: 10.1016/j.foodchem.2006.05.057

[19] Nachabe R et al. Diagnosis of breast cancer using diffuse optical spectroscopy from 500 to 1600 nm: Comparison of classification methods. Journal of Biomedical Optics. 2011;**16**(8):087010. DOI: 10.1117/1.3611010

[20] Li Q, He X, Wang Y, Liu H, Xu D, Guo F. Review of spectral imaging technology in biomedical engineering: Achievements and challenges. Journal of Biomedical Optics. 2013;**18**(10):100901. DOI: 10.1117/1.JBO.18.10.100901

[21] Wold JP et al. Non-contact Transflectance near infrared imaging for representative on-line sampling of dried salted coalfish (Bacalao). Journal of Near Infrared Spectroscopy. 2006;**14**(1): 59-66. DOI: 10.1255/jnirs.587

[22] Wu D, Sun D-W. Advanced applications of hyperspectral imaging technology for food quality and safety analysis and assessment: A review — Part I: Fundamentals. Innovative Food Science and Emerging Technologies. 2013;**19**:1-14. DOI: 10.1016/j.ifset.2013.04.014

[23] Sivertsen AH, Chu C-K, Wang L-C, Godtliebsen F, Heia K, Nilsen H. Ridge detection with application to automatic fish fillet inspection. Journal of Food Engineering. 2009;**90**(3):317-324. DOI: 10.1016/j.jfoodeng.2008.06.035

[24] Downey G. Non-invasive and non-destructive percutaneous analysis of farmed salmon flesh by near infra-red spectroscopy. Food Chemistry. 1996; **55**(3):305-311. DOI: 10.1016/0308-8146(95)00118-2

[25] Hsin-I Feng C. The tale of sushi: History and regulations. Comprehensive Reviews in Food Science and Food Safety. 2012;**11**(2):205-220. DOI: 10.1111/j.1541-4337.2011.00180.x

[26] Richardsen R and Østli J. Norwegian Trout in Japan. Consumer Preferences, Perceptions and Competitors. 2003.

[Online]. Available: https://nofima.brage.unit.no/nofima-xmlui/handle/11250/282799

[27] Zhu F, Zhang H, Shao Y, He Y, Ngadi M. Mapping of fat and moisture distribution in Atlantic Salmon using near-infrared hyperspectral imaging. Food and Bioprocess Technology. 2014; **7**(4):1208-1214. DOI: 10.1007/s11947-013-1228-z

[28] Zhang H et al. Non-destructive determination of fat and moisture contents in Salmon (Salmo salar) fillets using near-infrared hyperspectral imaging coupled with spectral and textural features. Journal of Food Composition and Analysis. 2020;**92**: 103567. DOI: 10.1016/j.jfca.2020.103567

[29] Dixit Y, Reis MM. Hyperspectral imaging for assessment of total fat in salmon fillets: A comparison between benchtop and snapshot systems. Journal of Food Engineering. 2023;**336**:111212. DOI: 10.1016/j.jfoodeng.2022.111212

[30] Skjelvareid MH, Heia K, Olsen SH, Stormo SK. Detection of blood in fish muscle by constrained spectral unmixing of hyperspectral images. Journal of Food Engineering. 2017;**212**:252-261. DOI: 10.1016/j.jfoodeng.2017.05.029

[31] Heia K, Washburn KE, and Skjelvareid MH. Automatic Quality Control of Internal Defects in cod - Results from Hyperspectral, Ultrasound and X-ray Imaging. 2017. [Online]. Available: https://nofima.brage.unit.no/nofima-xmlui/handle/11250/2480578

[32] Maritech AS. Maritech Eye. 2022. https://maritech.com/our-solutions-seafood-production/maritech-eye/

[33] Chen Z, Wang Q, Zhang H, Nie P. Hyperspectral imaging (HSI) Technology for the non-Destructive Freshness Assessment of pearl gentian grouper under different storage conditions. Sensors. 2021;**21**(2):583. DOI: 10.3390/s21020583

[34] Kimiya T, Sivertsen AH, Heia K. VIS/NIR spectroscopy for non-destructive freshness assessment of Atlantic salmon (Salmo salar L.) fillets. Journal of Food Engineering. 2013; **116**(3):758-764. DOI: 10.1016/j.jfoodeng.2013.01.008

[35] Sivertsen AH, Kimiya T, Heia K. Automatic freshness assessment of cod (Gadus morhua) fillets by Vis/Nir spectroscopy. Journal of Food Engineering. 2011;**103**(3):317-323. DOI: 10.1016/j.jfoodeng.2010.10.030

[36] Bekhit AE-DA, Holman BWB, Giteru SG, Hopkins DL. Total volatile basic nitrogen (TVB-N) and its role in meat spoilage: A review. Trends in Food Science and Technology. 2021;**109**: 280-302. DOI: 10.1016/j.tifs.2021.01.006

[37] Moosavi-Nasab M, Khoshnoudi-Nia S, Azimifar Z, Kamyab S. Evaluation of the total volatile basic nitrogen (TVB-N) content in fish fillets using hyperspectral imaging coupled with deep learning neural network and meta-analysis. Scientific Reports. 2021;**11**(1):5094. DOI: 10.1038/s41598-021-84659-y

[38] Cheng J-H, Sun D-W, Zeng X-A, Pu H-B. Non-destructive and rapid determination of TVB-N content for freshness evaluation of grass carp (Ctenopharyngodon idella) by hyperspectral imaging. Innovative Food Science and Emerging Technologies. 2014;**21**:179-187. DOI: 10.1016/j.ifset.2013.10.013

[39] Cheng J-H, Sun D-W, Wei Q. Enhancing visible and near-infrared hyperspectral imaging prediction of

TVB-N level for fish fillet freshness evaluation by filtering optimal variables. Food Analytical Methods. 2017;**10**(6): 1888-1898. DOI: 10.1007/s12161-016-0742-9

[40] Yu H-D et al. Hyperspectral imaging in combination with data fusion for rapid evaluation of tilapia fillet freshness. Food Chemistry. 2021;**348**: 129129. DOI: 10.1016/j.foodchem.2021.129129

[41] Zhang W, Cao A, Shi P, Cai L. Rapid evaluation of freshness of largemouth bass under different thawing methods using hyperspectral imaging. Food Control. 2021;**125**:108023. DOI: 10.1016/j.foodcont.2021.108023

[42] Khoshnoudi-Nia S, Moosavi-Nasab M. Prediction of various freshness indicators in fish fillets by one multispectral imaging system. Scientific Reports. 2019;**9**(1):14704. DOI: 10.1038/s41598-019-51264-z

[43] Wang QX, Su LH, Zou J, Chen NX, Wu T, Yang L. Research on hardness detection method of crisped grass carp based on visible - near infrared hyperspectral technology. Journal of Physics Conference Series. 2021;**1757**(1): 012002. DOI: 10.1088/1742-6596/1757/1/012002

[44] Khoshtaghaza MH, Khojastehnazhand M, Mojaradi B, Goodarzi M, Saeys W. Texture quality analysis of rainbow trout using hyperspectral imaging method. International Journal of Food Properties. 2016;**19**(5):974-983. DOI: 10.1080/10942912.2015.1042111

[45] Wang X, Shan J, Han S, Zhao J, Zhang Y. Optimization of fish quality by evaluation of Total volatile basic nitrogen (TVB-N) and texture profile analysis (TPA) by near-infrared (NIR) hyperspectral imaging. Analytical Letters. 2019;**52**(12):1845-1859. DOI: 10.1080/00032719.2019.1571077

[46] Wang S, Das AK, Pang J, Liang P. Real-time monitoring the color changes of large yellow croaker (Larimichthys crocea) fillets based on hyperspectral imaging empowered with artificial intelligence. Food Chemistry. 2022;**382**: 132343. DOI: 10.1016/j.foodchem.2022.132343

[47] McClelland G. Spatial and temporal distributions of larval sealworm (Pseudoterranova decipiens, Nematoda: Anisakinae), in Hippoglossoides platessoides (Pleuronectidae) in eastern Canada from 1980 to 1990. ICES Journal of Marine Science. 2000;**57**(1):69-88. DOI: 10.1006/jmsc.1999.0518

[48] Hafsteinsson H, Rizvi SSH. A review of the Sealworm problem: Biology, implications and solutions. Journal of Food Protection. 1987;**50**(1):70-84. DOI: 10.4315/0362-028X-50.1.70

[49] Mercken E et al. Sensitivity of candling as routine method for the detection and recovery of ascaridoids in commercial fish fillets. Scientific Reports. 2022;**12**(1):1358. DOI: 10.1038/s41598-022-05235-6

[50] Bublitz CG, Choudhury GS. Effect of light intensity and color on worker productivity and parasite detection efficiency during candling of cod fillets. Journal of Aquatic Food Product Technology. 1993;**1**(2):75-89. DOI: 10.1300/J030v01n02_08

[51] Pippy JHC. Use of ultraviolet light to find parasitic nematodes in situ. Journal of the Fisheries Board of Canada. 1970; **27**(5):963-965. DOI: 10.1139/f70-107

[52] Hafsteinsson H, Parker K, Chivers R, Rizvi SSH. Application of ultrasonic

waves to detect Sealworms in fish tissue. Journal of Food Science. 1989;**54**(2): 244-247. DOI: 10.1111/j.1365-2621.1989.tb03053.x

[53] Wold JP, Westad F, Heia K. Detection of parasites in cod fillets by using SIMCA classification in multispectral images in the visible and NIR region. Applied Spectroscopy. 2001; **55**(8):1025-1034. DOI: 10.1366/0003702011952929

[54] Pau LF. Fish Quality Control by Computer Vision. Abingdon, Oxfordshire: Routledge; 2017

[55] Stormo SK, Ernstsen A, Nilsen H, Heia K, Sivertsen AH, Elvevoll E. Compounds of parasitic roundworm absorbing in the visible region: Target molecules for detection of roundworm in Atlantic cod. Journal of Food Protection. 2004;**67**(7):1522-1525. DOI: 10.4315/0362-028X-67.7.1522

[56] Stormo SK, Sivertsen AH, Heia K, Nilsen H, Elvevoll E. Effects of single wavelength selection for Anisakid roundworm larvae detection through multispectral imaging. Journal of Food Protection. 2007;**70**(8):1890-1895. DOI: 10.4315/0362-028X-70.8.1890

[57] Sigernes F, Lorentzen DA, Heia K, Svenøe T. Multipurpose spectral imager. Applied Optics. 2000;**39**(18):3143. DOI: 10.1364/AO.39.003143

[58] Heia K, Sivertsen AH, Stormo SK, Elvevoll E, Wold JP, Nilsen H. Detection of nematodes in cod (Gadus morhua) fillets by imaging spectroscopy. Journal of Food Science. 2007;**72**(1):E011-E015. DOI: 10.1111/j.1750-3841.2006.00212.x

[59] Sivertsen AH, Heia K, Stormo SK, Elvevoll E, Nilsen H. Automatic nematode detection in cod fillets (Gadus Morhua) by Transillumination hyperspectral imaging. Journal of Food Science. 2011;**76**(1):S77-S83. DOI: 10.1111/j.1750-3841.2010.01928.x

[60] Sivertsen AH, Heia K, Hindberg K, Godtliebsen F. Automatic nematode detection in cod fillets (Gadus morhua L.) by hyperspectral imaging. Journal of Food Engineering. 2012;**111**(4):675-681. DOI: 10.1016/j.jfoodeng.2012.02.036

[61] Hassoun A et al. Fraud in animal origin food products: Advances in emerging spectroscopic detection methods over the past five years. Food. 2020;**9**(8):1069. DOI: 10.3390/foods9081069

[62] Chauvin J et al. Simulated annealing-based hyperspectral data optimization for fish species classification: Can the number of measured wavelengths Be reduced? Applied Sciences. 2021;**11**(22): 10628. DOI: 10.3390/app112210628

[63] Qin J et al. Detection of fish fillet substitution and mislabeling using multimode hyperspectral imaging techniques. Food Control. 2020;**114**: 107234. DOI: 10.1016/j.foodcont.2020.107234

[64] Kolmann MA et al. Hyperspectral data as a biodiversity screening tool can differentiate among diverse Neotropical fishes. Scientific Reports. 2021;**11**(1): 16157. DOI: 10.1038/s41598-021-95713-0

[65] Wang H, Qiu X, Zeng F, Shao W, Ma Q, Li M. Detection of physical descaling damage in carp based on hyperspectral images and dimension reduction of principal component analysis combined with pixel values. Journal of Food Science. 2022;**87**(6): 2663-2677. DOI: 10.1111/1750-3841.16144

[66] Jensen TK et al. Effect of the T90-codend on the catch quality of cod

(Gadus morhua) compared to the conventional codend configuration in the Barents Sea bottom trawl fishery. Fisheries Research. 2022;**250**:106277. DOI: 10.1016/j.fishres.2022.106277

[67] Jovanović B. Ingestion of microplastics by fish and its potential consequences from a physical perspective. Integrated Environmental Assessment and Management. 2017; **13**(3):510-515. DOI: 10.1002/ieam.1913

[68] Wootton N, Reis-Santos P, Gillanders BM. Microplastic in fish – A global synthesis. Reviews in Fish Biology and Fisheries. 2021;**31**(4):753-771. DOI: 10.1007/s11160-021-09684-6

[69] Zhang Y et al. Hyperspectral imaging based method for rapid detection of microplastics in the intestinal tracts of fish. Environmental Science & Technology. 2019;**53**(9): 5151-5158. DOI: 10.1021/acs.est.8b07321

[70] He H-J, Sun D-W. Selection of informative spectral wavelength for evaluating and Visualising Enterobacteriaceae contamination of Salmon flesh. Food Analytical Methods. 2015;**8**(10):2427-2436. DOI: 10.1007/ s12161-015-0122-x

[71] Naylor RL et al. A 20-year retrospective review of global aquaculture. Nature. 2021;**591**(7851): 551-563. DOI: 10.1038/s41586-021-03308-6

[72] Yue K, Shen Y. An overview of disruptive technologies for aquaculture. Aquaculture and Fisheries. 2022;**7**(2): 111-120. DOI: 10.1016/j.aaf.2021.04.009

[73] Svendsen E et al. Identification of spectral signature for in situ real-time monitoring of smoltification. Applied Optics. 2021;**60**(14):4127. DOI: 10.1364/ AO.420347

[74] Pettersen R, Lein Braa H, Gawel BA, Letnes PA, Sæther K, Aas LMS. Detection and classification of Lepeophterius salmonis (Krøyer, 1837) using underwater hyperspectral imaging. Aquacultural Engineering. 2019;**87**: 102025. DOI: 10.1016/j.aquaeng.2019. 102025

[75] Kong D et al. Hyperspectral imaging coupled with CNN: A powerful approach for quantitative identification of feather meal and fish by-product meal adulterated in marine fishmeal. Microchemical Journal. 2022;**180**:107517. DOI: 10.1016/j.microc.2022.107517

[76] Jaiswal G, Sharma A, Yadav SK. Critical insights into modern hyperspectral image applications through deep learning. WIREs: Data Mining and Knowledge Discovery. 2021; **11**(6):e1426. DOI: 10.1002/widm.1426

[77] Audebert N, Le Saux B, Lefevre S. Deep learning for classification of hyperspectral data: A comparative review. IEEE Geoscience Remote Sensor Management. 2019;**7**(2):159-173. DOI: 10.1109/MGRS.2019.2912563

[78] Wold JP, Veiseth-Kent E, Høst V, Løvland A. Rapid on-line detection and grading of wooden breast myopathy in chicken fillets by near-infrared spectroscopy. PLoS One. 2017;**12**(3): e0173384. DOI: 10.1371/journal. pone.0173384

[79] Wold JP, Kermit M, Woll A. Rapid nondestructive determination of edible meat content in crabs (cancer Pagurus) by near-infrared imaging spectroscopy. Applied Spectroscopy. 2010;**64**(7): 691-699. DOI: 10.1366/ 000370210791666273

[80] O'Farrell M, Wold JP, Høy M, Tschudi J, Schulerud H. On-line fat

content classification of inhomogeneous pork trimmings using multispectral near infrared Interactance imaging. Journal of Near Infrared Spectroscopy. 2010;**18**(2): 135-145. DOI: 10.1255/jnirs.876

[81] Wold JP, Solberg LE, Gaarder MØ, Carlehøg M, Sanden KW, Rødbotten R. In-line estimation of fat marbling in whole beef striploins (longissimus lumborum) by NIR hyperspectral imaging. A closer look at the role of myoglobin. Food. 2022;**11**(9):1219. DOI: 10.3390/foods11091219

[82] Wold JP, O'Farrell M, Høy M, Tschudi J. On-line determination and control of fat content in batches of beef trimmings by NIR imaging spectroscopy. Meat Science. 2011;**89**(3):317-324. DOI: 10.1016/j.meatsci.2011.05.001

Section 2

Preprocessing and Feature Extraction of Hyperspectral Imaging Data for Machine-Learning and Deep-Learning Analysis

Chapter 2

Useful Feature Extraction and Machine Learning Techniques for Identifying Unique Pattern Signatures Present in Hyperspectral Image Data

Jeanette Hariharan, Yiannis Ampatzidis, Jaafar Abdulridha and Ozgur Batuman

Abstract

This chapter introduces several feature extraction techniques (FETs) and machine learning algorithms (MLA) that are useful for pattern recognition in hyperspectral data analysis (HDA). This chapter provides a handbook of the most popular FETs that have proven successful. Machine learning algorithms (MLA) for use with HDA are becoming prevalent in pattern recognition literature. Several of these algorithms are explained in detail to provide the user with insights into applying these for pattern recognition. Unsupervised learning applications are useful when the system is provided with the correct set of independent variables. Various forms of linear regression assay adequately solve hyperspectral pattern resolution for identifying phenotypes. K-means is an unsupervised learning algorithm that is used for systematically dividing a dataset into K number of pattern groups. Supervised and unsupervised neural networks (NNs) are used to discern patterns in hyperspectral data with features as inputs and in large datasets where little *a priori* knowledge is applied. Other supervised machine learning procedures derive valuable feature detectors and descriptors through support vector machine. Several methods using reduced sets for extracting patterns from hyperspectral data are shown by discretized numerical techniques and transformation processes. The accuracy of these methods and their usefulness is generally assessed.

Keywords: pattern signature, hyperspectral data, data reduction, power spectral density, biomarker

1. Introduction

Hyperspectral imaging and data analysis have recently received considerable attention since the representative data is ultra-high resolution and informative [1–7].

Hyperspectral data is collected by using a precision imaging device which emits light energy at wavelengths below, within, and above the visible range. Other cameras such as RGB (Red-Green-Blue sensitive filters) and multispectral, have more limited sets of data. These passive sensors only collect one to five reflective values from ambient light that is present (**Figure 1**). It then scans each pixel location sequentially for reflectance values from each wavelength of light emitted. Each wavelength is spaced about 2–5 nms. Apart. The data collected for each pixel in an image then represents the complete spectrum returns for the hyperspectral bands presented (see **Figures 2–4**). Within this data, patterns of information exist that have never been detected before, thus allowing the explorer to glean relevant and new highlights from this collected data set.

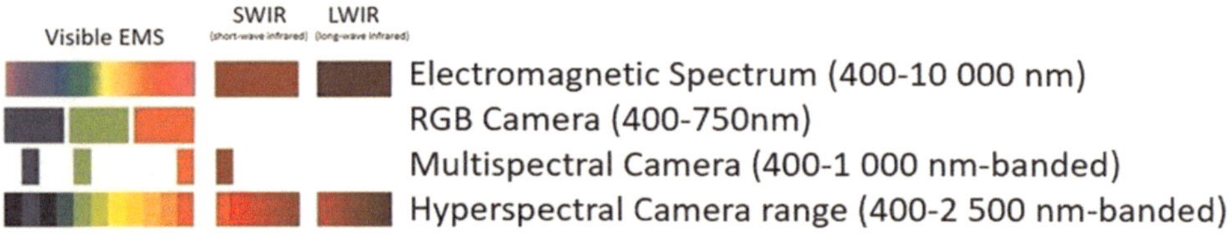

Figure 1.
Various camera operational bands.

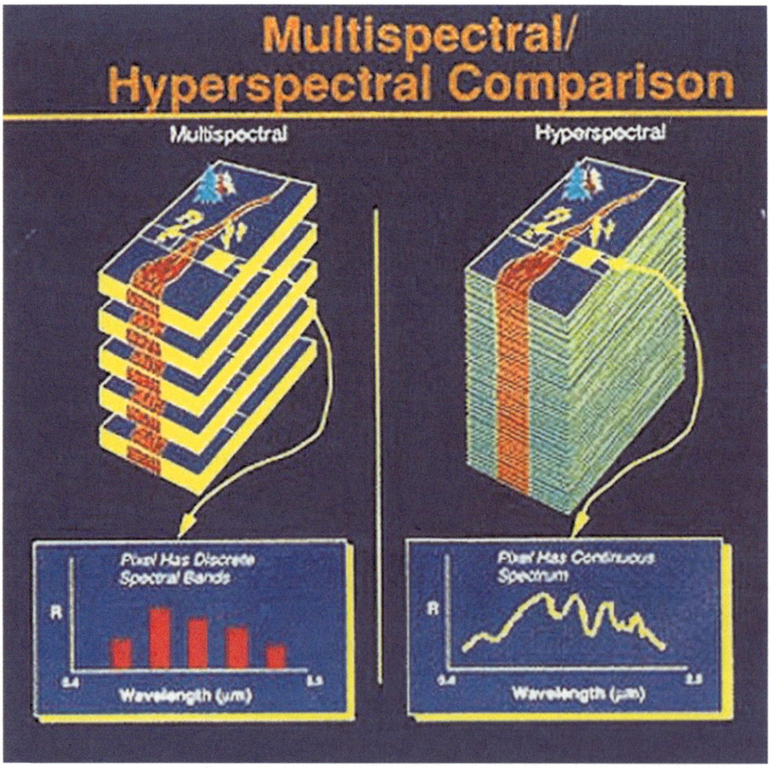

Figure 2.
Multispectral comparison with hyperspectral (Dr. Nicholas M. Short, Sr. – Public domain).

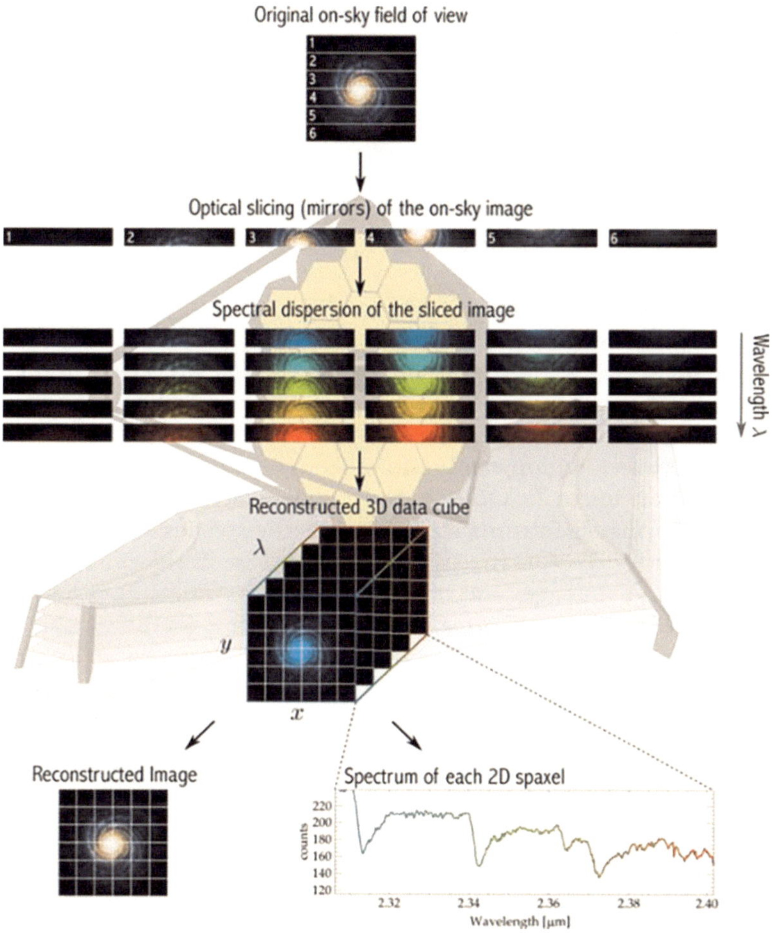

Figure 3.
Hyperspectral imaging process (NASA – Public domain).

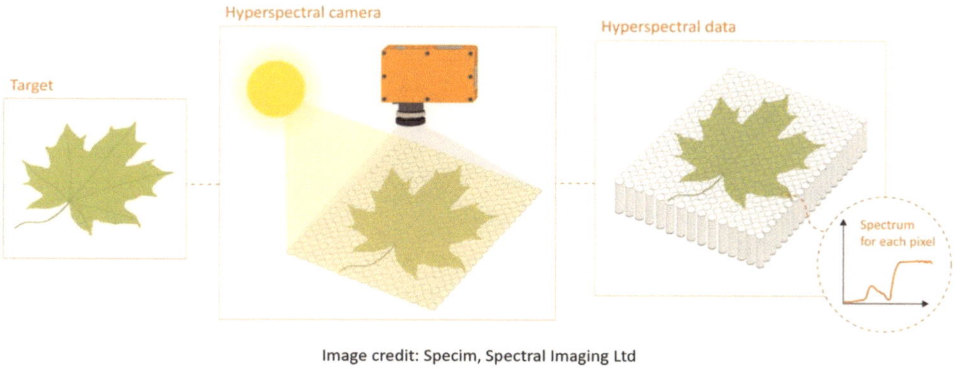

Figure 4.
Hyperspectral imaging analysis.

Hyperspectral cameras use a line scanning sensor (mostly, push broom type), that emits light at varying frequencies and then collects the reflected signal through a narrow slit. The narrower the slit, the higher the resolution of the camera, until it begins interfering with the light wave signal itself. The reflected light enters the slit and coincides with a concave mirror (**Figure 5**, M1) where the light is collimated. M1 redirects the collimated light from the scan to the optical grating. Here the light is divided or dispersed into its component frequencies. M2 acts to expand the beams and redirect the light to a reimaging lens array in the sensory unit.

The hyperspectral camera can be embedded on UAVs (e.g., **Figure 6**) to enhance aerial views for many image pursuits for agriculture, marine studies, search and rescue, surveillance, military activities, and construction site safety and management. Once hyperspectral images are stored, the data can be acquired per pixel per wavelength to reconstruct the image or study the reflected signatures. For instance, in agriculture, crops in a region can be surveyed by studying the map of the pure signature spectrum (**Figure 7**). Other pattern recognition algorithms can be used to understand how normal spectrums represent specific species of plants. Detecting plant diseases and stress factors in early stages of disease development is essential for selective and effective management of crop production. Through other data analysis procedures, such as feature extraction, statistical prediction, and reduced signature spectrums, pinpointing where the spectrums differ between a range of normal signatures and abnormal feature spectrums will reveal variations in the species [3, 8, 9]. Realizing how these spectrums differ because of diseases, abiotic stressors, nutritional deficiencies, or other factors, will give more useful information to the farmers.

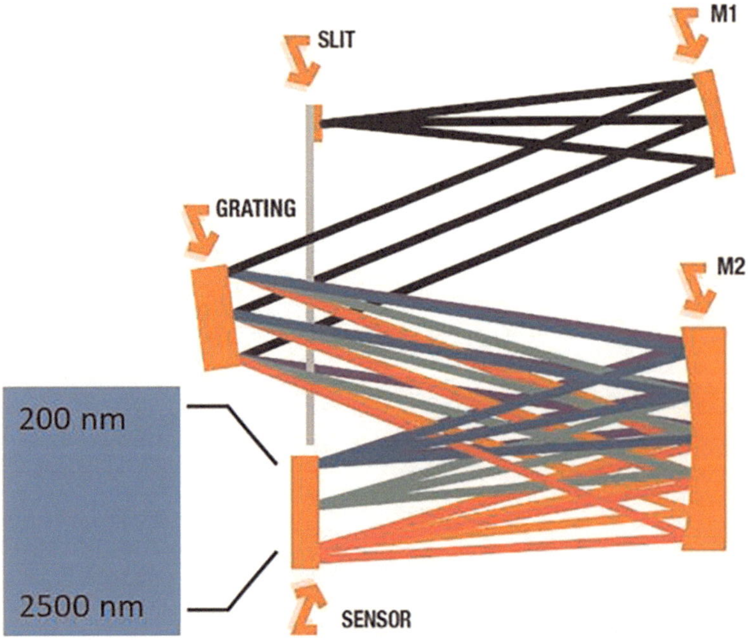

Figure 5.
Hyperspectral imaging hardware operation.

Figure 6.
Hyperspectral imaging device as payload on UAV.

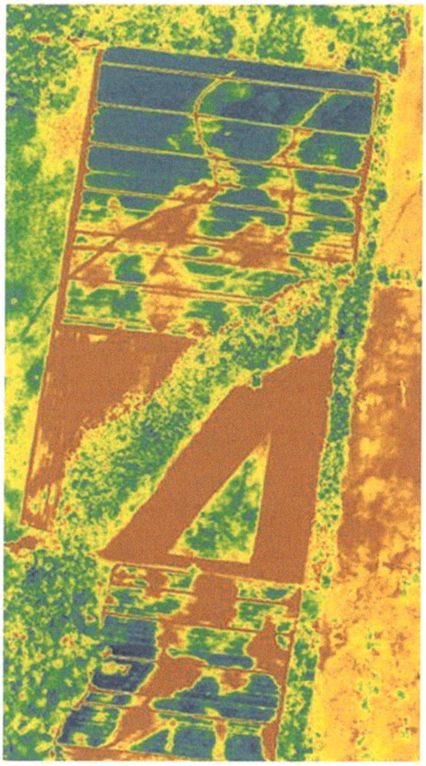

Figure 7.
Hyperspectral imaging of agricultural landscape (image credit: Geospatialworld.net).

The focus of this chapter is on extrication of features and pattern recognition algorithms that can be used in hyperspectral data analysis to obtain useful information. Common preprocessing and analysis applications include normalization and derivative spectra enhancement using finite differencing [10], complex step derivative [11], and derivative spectral shape equation [9]. Wavelet Transform has been used and compared to derivative spectra enhancement and shown to be very successful in spectral regions of interest; it is becoming more commonly used as an alternative to spectral derivative methods [12]. Polynomial interpolations are also used to smooth the (spectral) data and better represent enhanced spectra. Multivariate analysis can be used to gain a better understanding of spectral variance between feature data [2, 4]. Recently, autonomous ground and unmanned aerial vehicles with hyperspectral camera payloads have been used to collect data for agricultural purposes [13, 14]. Along with this method of data collection, deep and transfer learning artificial intelligence applications have been developed for pest and plant stress detection [5, 6]. These techniques required a high-quality training dataset for accurate development of the prediction models [5].

Hyperspectral Imaging is gaining widespread use in drone applications for agriculture and water safety. Agricultural applications include landscaping crop regions, analysis of crop health, understanding nutritional status of plants, harvest studies, flowering index, growth cycles comparison, trait discrimination, breeding information, and soil performance. Associated AI and machine learning applications are the mainstay of these informational systems. In water quality analysis, various hyperspectral algorithms such as partial least-squares, fully connected neural networks with backpropagation (FCNN-BP), Support vector machine (SVM) and Random Forest (RF) procedures have been successfully used and compared for quantitative investigation [15]. Other assessments have been implemented using FETs and AI for detecting water contamination in rivers [16], forest fire assessment, and automated drone team hyperspectral fusion.

Machine learning algorithms usually require some preprocessing of the data. Segmentation and feature extraction often use spatial filters, Laplacian of Gaussian with orientation filters [17], and other traditional methods that detect spectral discontinuities or similarities to adequately obtain prediction models for patterns. Gradient magnitude algorithms form ridges at high valued pixels, noted as watersheds, that are used to segment regions. Adaptive thresholding is applied to image data that has nonuniform background. This thresholding approach calculates local thresholds based on specified properties of pixel neighborhoods to segment regions of interest.

Active contours are another avenue of interest in obtaining features. Snakes were developed [18] as parametric curves that uncover boundary regions by minimizing an energy function. This optimization works well with hyperspectral data since it is relational to the spectra and energy reflected. Level sets use iterative solutions to find intersecting boundaries between features by optimizing a formulated level set equation. Using active contours, segmented bounded regions of interest can be brought to the forefront and presented as features in ML algorithms.

Features can be categorized as being primarily applicable to boundaries, regions, coded areas, spectral features or whole images. These are not mutually exclusive and can be used as feature map sections for convolutional neural networks. Hyperspectral signatures of various region features are often used to train neural networks and as inputs to MLA. Features in unsupervised learning environments are realized by the system.

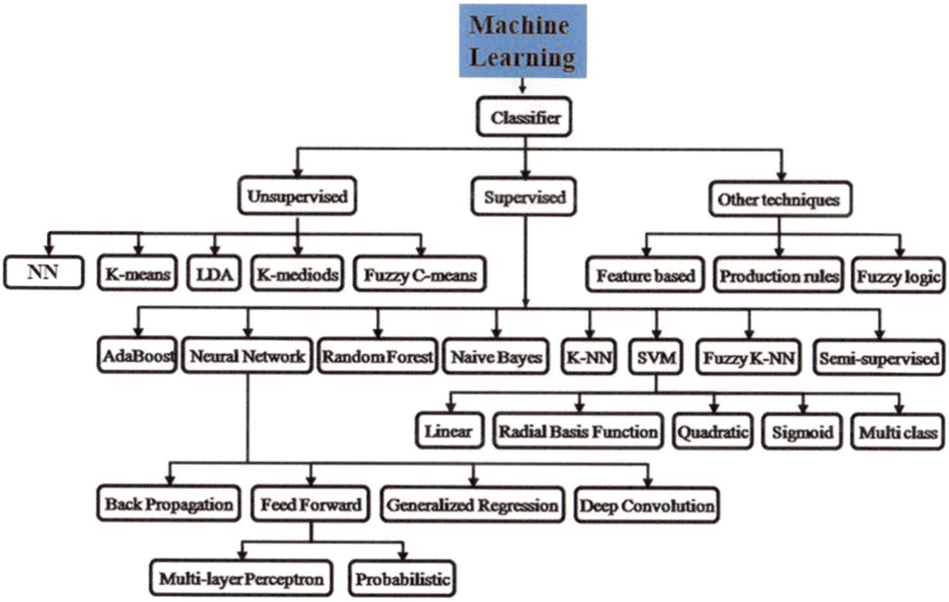

Figure 8.
Machine learning algorithms: Categories of methods to be considered when applying ML to hyperspectral data enhancement of soil data intensity range.

A branched diagram of popular MLAs is given by **Figure 8**. Supervised and unsupervised methods have their own distinct advantages and are dependent on the context of the application. This chapter reviews some useful methods in both categories and clarifies some of the subtle differences between these two types of algorithms. Other useful techniques for MLAs, such as fuzzy logic and quadratic nonlinear methods are depicted in the diagram of **Figure 8**. The reader is encouraged to explore these other methods and compare and contrast how these techniques can be used efficiently to enhance machine learning purposes.

Detecting plant diseases and stress factors in early stages of disease development is essential for selective and effective management of crop production. Laboratory analysis of plant samples for disease detection is time-consuming and labor-intensive. For that reason, several disease detection methods have been developed utilizing advanced and sophisticated hyperspectral data analysis approaches [3, 8, 9] and MLAs. These unique applications will be reviewed.

The rest of the chapter is divided into sections for:

i. Preprocessing methods that are used in hyperspectral data analysis

ii. FETs for supervised MLAs

iii. FETs for unsupervised machine learning

iv. Innovative methods for hyperspectral signature analysis in agriculture

v. Best practices for working with hyperspectral data and machine learning

2. Preprocessing methods for hyperspectral data analysis

Preprocessing of hyperspectral data involves numerical and statistical methods to filter noise and vibration more generally, as well as radiative transfer and empirical models used for airborne applications [19]. Some preprocessing methods include normalization of data, data smoothing, intensity transformation, histogram matching and histogram equalization, adaptive histogram equalization, correlation and convolution using spectral or spatial filters, and the use of fuzzy sets. This section reviews the most common and useful preprocessing methods for general purpose applications in hyperspectral data processing.

2.1 Normalization of hyperspectral data

Normalizing data taken from using similar sensors and similar methods is a common practice for purposes of understanding and clarifying data. This ensures the integrity of data presented for analysis, prediction and classification. It also provides some smoothing of the data using a standard normal variate transformation. The Standard Normal Variate (SNV) transformation is counter the effects of skewness of the data related to the reflectance spectra. The SNV was found to reduce error in the approximation which could be due to interferences caused by scattering and particle size differences. The probability distribution function that can be used for this standard normal variance transformation is given by

$$f(x) = \frac{1}{\sigma\sqrt{2\pi}} e^{-\zeta} \tag{1}$$

Where

$$\zeta = \frac{(x-\mu)^2}{2\sigma^2}$$

represents the SNV with mean, $\mu = 0$, and standard deviation, $\sigma = 1$.

Other distributions can be used for normalizing the data, if the camera manufacturer recommends or if there are other anomalies in the data that need to be taken into consideration. For example, if the data is non-Gaussian, the Z-score standardization can be used to force the data into standard normal distribution. Z-score standardization involves using the data mean and standard deviation to adjust all level points in the data such that

$$x_{ii} = \frac{x_i - \mu}{\sigma} \tag{2}$$

Then thex_{ii} values can be used as normally distributed data.

2.2 Hyperspectral data smoothing

After transforming the data into the SNV domain, the next steps might include resolving the data by curve fitting to smooth the spectral data enough so as to be able to reliably calculate the finite difference approximations or other numerical analysis of the data. A convolution method, such as the Savitzky–Golay Filter (SGF, [20]) can be used to approximate the spectra of the data.

For each pixel location, i, of the data, a signal spectrum of the data can be smoothed by:

$$S_i = \sum_N c_N f_i + n_C \tag{3}$$

Where.
S_i = smoothed pixel value per wavelength.
N = pixel neighborhood.
c_N = Coefficient of curve fit.
f_i = pixel value at each wavelength
n_C = center pixel value per wavelength

Determining the coefficients can be done by a box filter, applying Gaussian smoothing for the data set, or other filtering methods. For the SGF, the data is point transformed by polynomial approximation and the coefficients are found by a least square fit.

SGF takes into consideration the order (n) of the polynomial to which the data is being fitted, and the size of the window (m) inscribing the real data points which are being incorporated for the smoothing at each data point.

The SGF conversion process at the pixel level can be described by:

$$y(i) = \sum_{n=0}^{k} a_n i^n |i| \leq m \tag{4}$$

Where
y(i) = transformed output of Savitzky–Golay filter
n = term number of polynomials
a_n = coefficients of polynomial
k =order of polynomial fit

To obtain the values of the $n + 1$ coefficients, a least squares criterion is used for solving Eq. (4). By taking the partial derivative of the wrt a_n and setting the result equal to zero to minimize the error:

$$\frac{\partial}{\partial a_n}\left[\sum_{i=-m}^{m} (y(i) - s_i)^2\right] = 0 \tag{5}$$

Using Eq. (4) at i = 0 the first term can be found. Then using back-substitution and the criterion Eq. (5), the higher coefficients can be found

An example (**Figure 9**) shows the smoothing effect of the SGF applied to a sample hyperspectral signal in real-time data.

2.3 Intensity transformations

Intensity transformations can be used for data that is skewed because of background or foreground attenuations or ambient light interference. Stretching or

Figure 10.
Biocrust data (left) before histogram equalization; (right) after histogram equalization; note enhancement of soil.

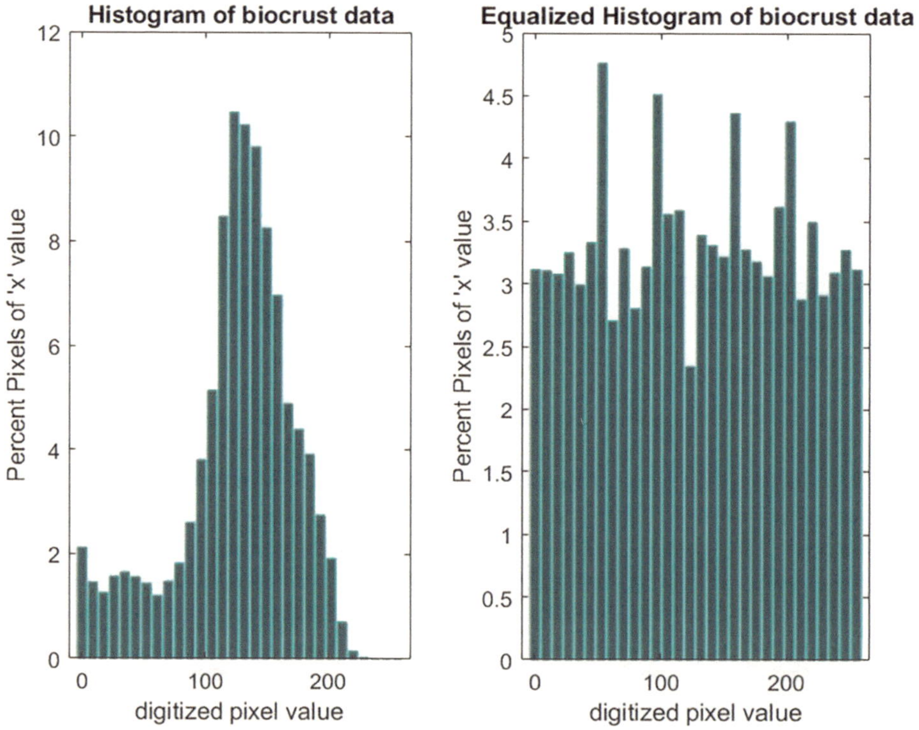

Figure 11.
Biocrust data (left) before histogram equalization; (right) after histogram equalization; note the broadening of the intensity range after equalization.

3. Feature extraction techniques for supervised machine learning

In order to find regions of interests or embedded patterns in the data, feature extraction methods for hyperspectral data is used to reduce the learning time and amount of data necessary for MLAs. When there is a priori knowledge in data, it is useful to extract this information so that the MLA used for pattern recognition is built on worthwhile information. In ordinary image data, finding lines, edges and corners is usually an advantageous effort since locating areas with sharp transitions is quite often associated with a pattern feature descriptor. With hyperspectral data, a more common application designed by Lowe [17] is scale-invariant feature transforms (SIFT).

There have been many adaptions of this SIFT transform over the years, but the robustness of this algorithm to find patterns in embedded frame data, real-time advancing frame data (such as moving target or moving platform), and for maximally stable extremal regions has become the supreme standard method used for keypoints feature extraction.

The first item that this transform addresses is scale invariance. By applying Gaussian filters to a stack of image(s) and increasing the smoothing by$k^n\sigma$ for each image that is stacked up to an octave (n = 4), these stacks of images are transformed by

$$\begin{aligned} \sigma &= \sigma_1 \\ \sigma_2 &= 2\sigma_1 \\ \sigma_n &= 2\sigma_{n-1} \end{aligned} \quad (14)$$

To find the keypoints, a difference operation is performed as given by Eq. (15):

$$\mathcal{L}(x, y, \sigma) = [g(x,\ y,\ k\sigma) - g(x,\ y,\ k\sigma)] \circledast \mathrm{I}(x,y) \quad (15)$$

Where
$\mathcal{L}(x, y, \sigma) = Laplacian\ operator$
$g(x, y, \sigma) = Gaussian\ transformed\ images$
I(x,y) = original image data

After this Laplacian of Gaussian operation is performed on the octave stacked smoothed image data, the extrema regions of the data begin to emerge. This scale-invariant region become the keypoint features of the data.

To build more robustness into this algorithm, invariance to rotation and other affine transformations are accounted for by applying orientation invariant gradient directional operators at the keypoints extracted from Eq. (13). These operators are 4^2 directional histogram matrices where each rotational element is 22.5 degrees differenced and weighted about bins that are multiples of 45°. After correlating this directional filter at the keypoints, keypoint descriptors are indicated and also used as feature directives for the keypoint features. With this collection of extrema data labeled as features, any number of machine learning methods can now be applied with the feature keypoints and descriptors provided as inputs.

The SIFT method was applied for soil biocrust data taken from US geological society biocrust data. Three band Electro-Optical (EO) imaging system - collected on June 2, 2018 using a Ricoh GR II camera (18.3 mm lens) mounted on 3DR Solo quad-rotor aerial vehicle (9:45 AM MDT) were collected [21].

The data before and after the SIFT procedure was applied is given in **Figure 12**. After applying a fully connected neural network with two layers, backpropagation and labeling the data, the supervised MLA was able to locate nine distinct areas of the terrain, including the two areas of biocrust.

4. A feature extraction technique for unsupervised machine learning

While many variations for linear discriminant analysis (LDA) exist, the focus on K–means and K-medoids has gotten less attention. The main emphasis of these methods is to use clustering of data traits to classify features. Clustering algorithms are vital knowledge acquisition tools [22]. Numerical clustering algorithms generally use a

References

[1] Abdulridha J, Batuman O, Ampatzidis Y. UAV-based remote sensing technique to detect Citrus canker disease utilizing hyperspectral imaging and machine learning. Remote Sensing. 2019;**11**(11):1373

[2] Abdulridha J, Ehsani R, Ampatzidis Y, de Castro A. Evaluating the performance of spectral features and multivariate analysis tools to detect Laurel wilt disease and nutritional deficiency in avocado. Computers and Electronics in Agriculture. 2018;**155**: 203-211

[3] Abdulridha J, Ampatzidis Y, Qureshi J, Roberts P. Identification and classification of downy mildew development stages in watermelon utilizing aerial and ground remote sensing and machine learning. Frontiers in Plant Science. 2022;**13**: 791018. DOI: 10.3389/fpls.2022.791018

[4] Ampatzidis Y, De Bellis L, Luvisi A. iPathology: Robotic applications and management of plants and plant diseases. Sustainability. 2017;**9**(6):1010. DOI: 10.3390/su9061010

[5] Ampatzidis Y, Partel V. UAV-based high throughput phenotyping in Citrus utilizing multispectral imaging and artificial intelligence. Remote Sensing. 2019;**11**(4):410. DOI: 10.3390/rs11040410

[6] Ampatzidis Y, Partel V, Costa L. Agroview: Cloud-based applications to process, analyze and visualize uAV-collected data for precision agriculture applications utilizing artificial intelligence. Computers and Electronics in Agriculture. 2020;**174**(July): 105157. DOI: 10.1016/j.compag.2020.105457

[7] Harakannanavar S, Rudagi J, Puranikmath V, Siddiqua A, Pramodhini R. Plant leaf disease detection using computer vision and machine learning algorithms. Global Transitions Proceedings. 2022;**3**: 305-310

[8] Hariharan J, Fuller J, Ampatzidis Y, Abdulridha J, Lerwill A. Finite difference analysis and bivariate correlation of hyperspectral data for detecting Laurel wilt disease and nutritional deficiency in avocado. Remote Sens. 2019;**11**(15):1748. DOI: 10.3390/rs11151748

[9] Paine E, Slonecker E, Simon N, Rosen B, Resmini R, Allen D. Optical characterization of two cyanobacteria genera, Aphanizomenon and Microcystis, with hyperspectral-microscopy. Journal of Applied Remote Sensing. 2018;**12**(3)

[10] Wang S, Celebi M, Zhang Y, Yu X, Lu S, Yao X, et al. Advances in data Preprocesing for biomedical data fusion: An overview of the methods, challenges, and prospects. Information Fusion. 2021; **76**:376-421

[11] Kiran R, Khandelwal K. Complex step derivative approximation for numerical evaluation of tangent moduli. Computers & Structures. 2014;**140**:1-13

[12] Susic N, Zibrat U, Sirca S, Strajnar P, Razinger J, Knapic M, et al. Discrimination between abiotic and biotic drought stress in tomatoes using hyperspectral imaging. Sensors and Actuators B: Chemical. 2018;**273**:842-852

[13] Poudyal C, Costa L, Sandhu H, Ampatzidis Y, Odero DC, Arbelo OC, et al. Sugarcane yield prediction and genotype selection using UAV-based

hyperspectral imaging and machine learning. Agronomy Journal. 2022:**114**: 2320–2333. DOI: 10.1002/agj2.21133

[14] Vijayakumar V, Ampatzidis Y, Costa L. Tree-level Citrus yield prediction utilizing ground and aerial machine vision and machine learning. Smart Agricultural Technology. 2022: 100077. DOI: 10.1016/j.atech.2022.100077

[15] Liu H, Yu T, Hu B, Hou X, Qian B. UAV-borne hyperspectral imaging remote sensing system based on Acousto-optic tunable filter for water quality monitoring. Remote Sensing. 2021;**13**(20):4069. DOI: 10.3390/rs13204069

[16] Scott N, Moore I. Nonnegative matrix factorization-based feature selection analysis for hyperspectral imagery of sediment-laden riverine flow. SPIE. 2018:1063114. DOI: 10.1117/12.2301273

[17] Lowe D. Distinctive image features from scale-invariance Keypoints. International Journal of Computer Vision. 2004;**60**(2):91–110. Corpus ID: 174065

[18] Kass M, Witkin A, Terzo Poulos D. Snakes: Active contour models. International Journal of Computer Vision. 1988:321-331

[19] Minu P, Lotliker A, Shaju S, Ashraf P, Kumar TS, Meenakumari B. Performance of operational satellite bio-optic algorithms in different water types in the southeastern Arabian Sea. Oceanologia. 2016;**58**: 317-313

[20] Jardim R, Morgado-Dias F. Savitzky-Golay filtering as image noise reduction with sharp color reset. Microprocessors and Microsystems. 2020;**74**:103006

[21] Savitzky A, Golay MJE. Smoothing and differentiation of data by simplified least squares procedures. Analytical Chemistry. 1964;**36**(8):1627-1639. DOI: 10.1021%2Fac60214a047

[22] Hartigan JA, Wong MA. Algorithm AS 136: A k-means clustering algorithm. Journal of the Royal Statistical Society, Series C. 1979;**28**(1):100-108 JSTOR 2346830

[23] Kaufman L, Rousseeuw PJ. Partitioning around Medoids (program PAM). In: Wiley Series in Probability and Statistics. Hoboken, NJ, USA: John Wiley & Sons, Inc.; 1990. pp. 68-125. DOI: 10.1002/9780470316801.ch2

[24] Fontanella L, Ippoliti L. Time series analysis: Methods and applications. Handbook of Statistics. 2012;**30**(17): 497-520

Section 3

Deep-Learning Enabling High-Resolution Hyperspectral Imaging

Chapter 3

Unsupervised Deep Hyperspectral Image Super-Resolution

Zhe Liu and Xian-Hua Han

Abstract

This chapter presents the recent advanced deep unsupervised hyperspectral (HS) image super-resolution framework for automatically generating a high-resolution (HR) HS image from its low-resolution (LR) HS and high-resolution RGB observations without any external sample. We incorporate the deep learned priors of the underlying structure in the latent HR-HS image with the mathematical model for formulating the degradation procedures of the observed LR-HS and HR-RGB observations and introduce an unsupervised end-to-end deep prior learning network for robust HR-HS image recovery. Experiments on two benchmark datasets validated that the proposed method manifest very impressive performance, and is even better than most state-of-the-art supervised learning approaches.

Keywords: deep learning, unsupervised learning, hyperspectral image, super-resolution, generative network

1. Introduction

Hyperspectral images (HSI) feature hundreds of bands with extensive spectral qualities that are helpful for a range of visual tasks, such as computer vision [1], mineral exploration [2], medical diagnosis [3], remote sensing [4], and so on. Due to technology restrictions, it is harder to capture high-quality HSI, and the acquired HSI has substantially lower resolution. As a result, super-resolution (SR) has been applied to obtain a HR-HSI, but it is a challenge because of texture blurring and spectral distortion problems at high magnifications. Thus, researchers frequently combine high-resolution PAN and low-resolution HSI [5] to achieve SR tasks. In recent years, it is a trend to fuse a high-resolution multispectral/RGB (HR-MS/RGB) image and a low-resolution hyperspectral (LR-HS) image for generating a high-resolution hyperspectral (HR-HS) image, which is called hyperspectral image super-resolution (HSI-SR). The HSI-SR methods are classified into two primary categories based on reconstruction principles: conventional mathematical model-based methods and deep learning-based approaches in a supervised/unsupervised manner. The following sections go into further information about each of these categories.

1.1 Mathematical model-based methods

Since HSI-SR is typically an inverse problem, a mathematical model-based approach yields a solution space that is far bigger than the actual result needed. In order to tackle this issue, mathematical model-based HSI-SR constrains the solution space using hand-crafted prior knowledge, regularizes the mathematical model, and then optimizes the model by minimizing the reconstruction errors. This method aims at establishing a mathematical formulation that simulates the transformation of HR-HS images into LR-HS and HR-RGB images. This process is extremely difficult, and direct optimization of the formed mathematical model might result in very unreliable solutions, as the known variables in the LR-HS/HR-RGB images under consideration are significantly less than the unknown variables to be estimated in the latent HR-HS images. In order to narrow the set of possible solutions, existing approaches often utilize a variety of priors to modify the mathematical model.

Based on prior knowledge of various structures, three categories of mathematical model-based HSI-SR methods can currently be distinguished: spectral unmixing-based methods [6], sparse representation-based methods [7], and tensor factorization-based methods [8]. For spectrum unmixing-based methods, Yokoya et al. [9] proposed a coupled non-negative matrix decomposition approach (CNMF), which alternatively unmixes LR-HS images and HR-RGB images to estimate HR-HS images. Later, Lanaras et al. [6] proposed a similar framework to jointly unmix the two input images by decoupling the initial optimization problem into two constrained least square problems. Dong et al. [7] incorporated alternating multiplication method (ADMM) techniques for solving the spectra unmixing model. Additionally, the sparse representation is frequently used as an alternative mathematical model for HSI-SR. In this model, the underlying HR-HS image is recovered by first learning the spectral dictionary from the LR-HS image under consideration, and then calculating the sparse coefficient of the HR-RGB image. Inspired by the existed spectral similarity of the neighboring pixels in the latent HS image, Akhtar et al. [10] proposed to perform group sparse and non-negativity representation within a small patch, while Kawakami et al. [11] applied a sparse regularizer for the decomposition of spectral dictionaries. Moreover, the tensor factorization-based method demonstrated that it could be used to resolve the HSI-SR problem. He et al. [8] factorized the HR-HS image into two low-rankness constraint matrices and achieved great super-resolution performances, which were motivated by the intrinsic low dimensionality of the spectrum space and the three-dimensional structure of the HR-HS image.

Despite some advancements in handcrafted prior, HSI-SR performance tends to be inconsistent and can cause severe spectral distortion due to the under-representation of handcrafted prior, depending on the content of the image under investigation.

1.2 Deep learning-based methods

Hyperspectral super-resolution is a hot field of research in hyperspectral imaging, as it can improve low-resolution images in both the spatial and spectral domains, turning them into high-resolution hyperspectral images. HSI-SR is a classic inverse problem, and deep learning has a lot of promise for resolving it. Depending on whether a training dataset is provided, supervised and unsupervised learning are the two approaches used in deep learning-based HSI-SR. A labeled training dataset is necessary for supervised learning in order to create a function or model from which subsequent data is fed in order to generate accurate predictions. But a labeled training dataset is not necessary for unsupervised learning.

1.2.1 Deep supervised learning-based methods

Different vision tasks have been successfully resolved by DCNNs. As a result, DCNN-based methods have been suggested for HSI-SR tasks, which eliminate the requirement to investigate various manually handcrafted priors. With the LR-HS observation only, Li et al. [12] presented an HSI-SR model by combining a spatial constraint (SCT) strategy with a deep spectral difference convolutional neural network (SDCNN). Han et al. [13] utilized three straightforward convolutional layers in the groundbreaking HS/RGB fusion work, whereas later work utilized more advanced CNN architectures, such as ResNet [14] and DenseNet [15], in an effort to attain more robust learning capabilities. By resolving the Sylvester equation using a fusion framework, Dian et al. [16] first provided an optimization technique, and then they investigated a DCNN-based strategy to enhance the initialization results. Further, Han et al. [17] proposed a multi-layer, multi-level spatial, and spectral fusion network that successfully fused existing LR-HS and HR-RGB images. In order to investigate an MS/HS fusion network and optimize the suggested MS/HS fusion system, Xie et al. [18] employed a low-resolution imaging model and spectral low-level knowledge of HR-HS images. In order to solve HS image reconstruction difficulties effectively and accurately, Zhu et al. [19] researched the progressive zero-centric residual network (PZRes-Net), a lightweight deep neural network-based system. All the DCNN-based methods mentioned above take training with a large number of pre-prepared training instances that contain not only LR-HS and HR-RGB images but also the corresponding HR-HS images as labels, that is, the set of training triples, despite the fact that the reconstruction performance was significantly improved.

1.2.2 Deep unsupervised learning-based methods

Although HS images are difficult to obtain in the real world, deep learning networks for HSI-SR require a lot of hyperspectral images as training data. It is rather challenging to collect good quality HSIs due to hardware restrictions, and the resolution of the acquired HSIs is relatively low. For supervised learning, which needs big training datasets to succeed, this is an unsolvable problem. As a result, unsupervised learning is one of the key research areas. Unlike supervised learning, unsupervised learning does not require any HR-HS image as a ground-truth image and uses only easily accessible HR-MS/RGB images and LR-HS images to generate HR-HS images.

It is well known that the corresponding training triplets, especially the HR-HS images, are extremely hard to be collected in real applications. Thus, the quality and amount of the collected training triplets generally become the bottleneck of the DCNN-based methods. Most recently, Qu et al. [20] attempted to solve the HSI super-resolution problem in an unsupervised way and designed an encoder-decoder architecture for exploiting the approximate low-rank prior structure of the spectral model in the latent HR-HS image. This unsupervised framework did not require any training samples in an HSI dataset and could restore the HR-HS image using a CNN-based end-to-end network. However, this method needed to be carefully optimized step-by-step in an alternating way, and the HS image recovery performance was still not enough. Liu et al. [21] proposed an unsupervised multispectral and hyperspectral image fusion (UnMHF) network using the observations of the under-studying scene only, which estimates the latent HR-HS image with the learned encoder-decoder-based generative network from a noise input and can only be adopted to the observed LR-HS and HR-RGB image with the known spatial downsampling operation and camera spectral function (CSF). Later, Uezato et al. [22] exploited a similar method for unsupervised image pair fusion, dubbed a guided deep decoder (GDD) network for the known spatial and spectral degradation operation only. Thus, the UnMHF [21] and GDD [22] can be categorized into the non-blind paradigm, and lack of generalization in a real scenario. Zhang et al. [23] proposed two steps of learning methods via modeling the common priors of the HR-HS image in a supervised way and then adapting to the under-studying scene for modeling it's specific prior in an unsupervised manner. In addition, the unsupervised adaptation is capable of learning the spatial degradation operation of the observed LR-HS image but can only deal with the observed HR-HS image with known CSF, and thus it would be categorized as a semi-blind paradigm for possibly learning the spatial degradation operations only in the observed LR-HS image. Moreover, Fu et al. [24] exploited an unsupervised hyperspectral image super-resolution method using the designed loss function formulated by the observed LR-HS and HR-RGB images only and integrated a CSR optimization layer after the HSI super-resolution network to automatically select or learn the optimal CSR for adapting to the target RGB image possibly captured by various color cameras, which is also divided into the semi-blind paradigm for possibly learning the spectral degradation operation: CSF only. Further, the unsupervised adaptation subnet in ref. [23] and the method [24] utilize the under-studying observed images only instead of the requirement of additional training samples for guiding the network training, which achieved impressive performance as an unsupervised learning strategy. However, these learning methods based on the under-studying observed images only are easy to drop into a local solution, and the final prediction heavily depends on the initial input of the network. Our method is also formulated in this unsupervised learning paradigm, and we are going to clarify the distinctiveness of our method in the next sub-section.

2. The proposed unsupervised learning-based methods

In this section, we first describe the problem formulation in the HSI-SR task and then present the proposed deep unsupervised learning-based method.

2.1 Problem formulation

Let us consider image pairs: a LR-HS image $X \in R^{w \times h \times L}$, where w and h are the width and height, and a HR-RGB image $Y \in R^{W \times H \times 3}$, where w and h are the width and height of Y and Z, respectively. A HR-HS image $Z \in R^{W \times H \times L}$, where L is the number of spectral channels in the HR-HS image, is what we are trying to reconstruct for HSI-SR. The following formula can be used to represent the degradation between the HR-HS target image and the observed images: X and Y.

$$X = \mathrm{k}^{(spa)} \otimes Z^{(spa)} \downarrow + n_x, Y = Z * C^{(Spec)} + n_y, \qquad (1)$$

where $\otimes$ stands for the convolution operator, $(\mathrm{Spa}) \downarrow$ for the spatial domain downsampling operator, and $\mathrm{k}^{(Spa)}$ for the two-dimensional blur kernel in the spatial domain. Three one-dimensional spectral filters $\mathrm{C}^{(Spec)}$ constitute the spectral sensitivity function of RGB cameras, which translates L spectral bands to RGB bands. The additive white Gaussian noise (AWGN) with noise level σ is represented by n_x and n_y. We rephrase the degenerate model as a matrix formulation to quantify the problem, that is,

$$X = DBZ + n_x, Y = ZC + n_y, \tag{2}$$

where B is the spatial blur matrix, D is the downsampling matrix, and C is the transformation matrix representing the spectral sensitivity function (CSF). According to Eqs. (1) and (2), a general HSI-SR task should evaluate $k^{(Spa)}$ (or B), $(Spa)\downarrow$ (or D), and $C^{(Spec)}$ (or C) from observed image pairs X and Y, which makes it very complicated to obtain the latent Z. It is a challenging problem that has rarely been studied in the HSI-SR task. Therefore, the general solution is to assume that the blur kernel type and spectral sensitivity function (CSF) of the RGB camera are known and to approximate them by some mathematical operations in the application. The current study followed to the previous setting in principle, but we also investigated whether it was possible to reconstruct HR-HS images without knowing the kind of CSF or the blur kernel beforehand as a generic solution for a specific scenario.

Let us begin by defining the generic formula of the HSI-SR task generally. The maximum a posterior (MAP) framework is the foundation formula of the majority of classical approaches.

$$Z^* = \underset{Z}{\text{argmax}}\Pr(Z|X, Y, B, C) = \underset{Z}{\text{argmax}}\Pr_{(B, C)}(X, Y|Z)\Pr(Z), \tag{3}$$

where $Pr(Z)$ performs prior modeling of latent HR-HS images and $\Pr_{(B,C)}(X,Y|Z)$ is the likelihood of the fidelity term corresponding to the known kernel type and CSF matrix. With regard to the latent HR-HS image Z, which we define as $-\log\left(\Pr_{(B, C)}(X, Y|Z)\right)$, it is specifically assumed that the reconstruction errors of the fidelity terms X and Y are independent Gaussian distribution in general. The prior modeling of HR-HS images is subjected to the regularization requirement $-\log(\Pr(Z)) = \phi(Z)$. The reconstruction model of the MAP-based HSI SR in Eq. (3) can be redefined using the following formula.

$$Z^* = \underset{Z}{\text{argmin}}\alpha\beta_1\|X - DBZ\|_F^2 + (1-\alpha)\beta_2\|Y - ZC\|_F^2 + \lambda\phi(Z), \tag{4}$$

where $\|\bullet\|_F$ represents the Frobenius norm. It is generally necessary to introduce normalization weights, such as $\beta_1 = 1/N_1$ and $\beta_2 = 1/N_2$, where N_1 and N_2 are multiples of the number of pixels and spectral bands in LR-HS and HR-HS images, respectively. This is because HR-RGB and LR-HS images have different numbers of elements. In addition, we further modify the contribution of these two reconstruction errors using the hyperparameter $\alpha(0 \le \alpha \le 1)$. On the other hand, the trade-off adjustment parameter is λ. We have experimentally developed appropriate prior parameters as regularization term $\phi(Z)$ in order to obtain a robust solution. Numerous prior restrictions have been present. The employed priors, however, are often manually determined and fall short of adequately describing the intricate structure of HR-HS images. Furthermore, the established priors should vary depending on the details of the situation being studied, and choosing the suitable priors for a specific scenario is still an art.

The DCNN method is one of the most recent deep learning-based HSI-SR techniques. It effectively captures prospective HS image features (common prior) in a fully supervised learning manner utilizing previously trained training samples (external datasets). Particularly supervised deep learning methods seek to learn joint CNN models by minimizing such loss functions given N trainable triples.

$$(X_i, Y_i, Z_i)(i = 1, 2, \cdots, N).$$

$$\theta^* = \underset{\theta}{\text{argmin}} \sum_{i}^{N} \|Z_i - F_{CNN}(X_i, Y_i)\|_F^2, \tag{5}$$

where F_{CNN} stands for a DCNN network transform with θ learning parameters. In contrast to directly searching in the ground-truth image space Z, these approaches are trained to extract the optimal parameters θ^* of the network, and they are able to identify common prior variables concealed in the training samples utilizing the powerful and effective DCNN modeling capabilities. The underlying HR-HS images for each observation (X_t,Y_t) in the research can be simply rebuilt as: $\hat{Z}_t = F_{CNN}^{\theta^*}(X_t, Y_t)$ after learning the network parameters θ^*. Although these supervised deep learning methods have shown encouraging results, it is necessary to provide a substantial training dataset that includes LR-HS, HR-RGB, and HR-HS images—all of which are particularly challenging to gather in HSI-SR tasks—in order to learn a good model.

2.2 The overview motivation

Recent deep learning-based HSI-SR techniques have demonstrated that DCNNs perform well and are capable of accurately capturing the underlying spatial and spectral structure (joint prior information) of potential HS images. The training labels (HR-HS images) for these algorithms, which are typically performed in a fully supervised way and need large-scale training datasets containing LR-HS, HR-RGB, and HR-HS images, are challenging to gather. Numerous studies on natural image generation (DCGAN [25]) and its variations have demonstrated that high-resolution, high-quality images with specific features and attributes can be produced from noisy random input data without the supervision of high-quality ground-truth data. This indicates that originating from a random initial image and scanning the parameter space of a neural network can capture the inherent structure (a prior) of possible images with certain features. DIPs [26] have also been utilized to properly perform a number of natural image restoration tasks, including image separation, blurring, and super-resolution extraction, using just the degraded version of a scene to guide them. This unsupervised paradigm is used in the current study, which tries to learn the precise spatial and spectral structure (a prior) of HR-HS latent images from degraded data (LR-HS and HR-RGB images).

The spatial and spectral structure of the underlying HR-HS image Z was specifically modeled using the generative neural network G_θ (θ is a network parameter that must be learnt). The fusion-based HSI-SR model can be rebuilt as follows by substituting Z with G_θ in Eq. (4) and deleting the regularization term $\phi(Z)$ connected with the prior acquired automatically by the generative network.

$$\theta^* = \underset{\theta}{\text{argmin}} \alpha\beta_1 \|X - DBG_\theta(Z_{in})\|_F^2 + (1-\alpha)\beta_2 \|Y - G_\theta(Z_{in})C\|_F^2, \tag{6}$$

where $G_\theta(Z_{\text{in}})_i$ is the *i-th* component of the HR-HS estimation and Z_{in} is the input to the generative neural network. Eq. (6) tries to explore the parameter space of the generative neural network G_θ by leveraging the powerful modeling capability to generate a more reliable HR-HS image, instead of directly searching the exceedingly vast, non-uniquely determined raw HR-HS space.

To solve the above unsupervised HSI-SR task, there are still several issues to be needed to elaborately address: (i) How to design the generative network's architecture so that both spectral correlations and low-level spatial statistics can be effectively modeled during training. (ii) What kind of input to the generative network should be employed so that the local minimization point can be avoided. (iii) How to implement an end-to-end learning framework for incorporating different degradation operations (blurring, downsampling, and spectral modification) following the generative network. In the next sections, we embody the solutions to the aforementioned issues.

2.3 Architecture of the generative neural network

Generative neural networks G_θ can be implemented using arbitrary DCNN architectures. A generative neural network G_θ is required to offer acceptable modeling skills due to the diversity of information, including potentially significant structures, rich textures, and complicated spectra in HR-HS images. It has been demonstrated that various generative neural networks have a great deal of promise for producing high-quality natural images [Pix2pix and others], for example, in adversarial learning settings [27]. In this study, a multi-level feature learning architecture is employed, along with simplified encoder-decoder features and an encoder-decoder architecture that allows for feature reuse via skip connections between the encoder and the decoder. **Figure 1** shows a thorough representation of a generative neural network.

Five blocks compensate the encoder and decoder, and they both learn representative features at various scales. To reuse the extracted detailed features, the output of each of the 5 encoder-side blocks is straight-through forwarded to the corresponding decoder. A maximum clustering layer with a 2 × 2 kernel is used to reduce the size of the feature map between encoder blocks, and an upconversion layer is used to double

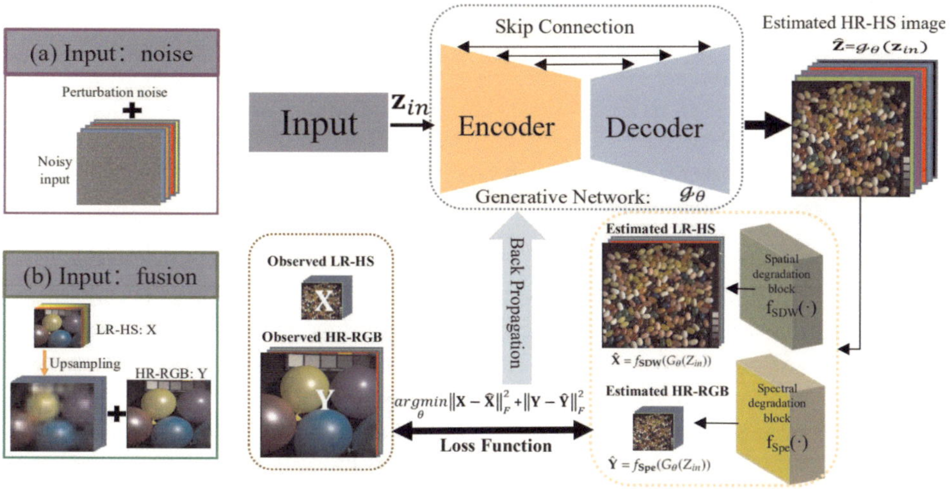

Figure 1.
Conceptual diagram of the proposed unsupervised deep HSI-SR.

the size of the feature map between decoder blocks for recovery. Each block is comprised of three convolutional layers that each follow the RELU activation function. Finally, the HR-HS images are estimated using the convolutional output layer. The training state of the generative neural network cannot be estimated or guided in an unsupervised learning environment as there is no ground-truth HR-HS image. The assessment criteria listed in Eq. (6) are then generated using the observed HR-RGB and LR-HS images.

2.4 Input data to the generative neural network

We classify the input data into two types. The first is a noisy input with a random perturbation added to check the robustness, corresponding to the deep unsupervised fusion learning (DUFL) model; in particular, to contrast with the addition of random perturbation, we also perform experiments without random perturbation, that is, the DUFL+ model. The second input data is the fusion context of fused observations HR-RGB and LR-HS, which corresponds to the deep self-supervised HS image reconstruction (DSSH) framework.

2.4.1 The noise input

The deep image prior network (DIP) [26] was developed to get low spatial statistics using inputs of uniformly distributed noise vectors generated at random. Nevertheless, because the noise vectors are chosen at random, DIP has a limited ability to discover spectral and spatial correlations and is more challenging to tune. Motivated by the DIP, we proposed a deep unsupervised fusion learning (DUFL) model, in which a common generative neural network is trained to generate target images with predetermined features; typically, a randomly selected noise vector based on a distribution function (for example, Gaussian or uniform distribution) is used as input to ensure that the generated images have enough diversity and variability. The observed degradation (LR-HS and HR-RGB images) of the corresponding HR-HS images is required for our HSI-SR task. Therefore, it makes sense to determine the best network parameter space for searching a given HR-HS image as the previously sampled noise vector Z_{in}^{0}. However, a constant noise input could lead to a local minimum in the generative neural network. As a result, the HR-HS image's estimate is inaccurate. Therefore, it is suggested to disturb the fixed initial input with a small randomly generated noise vector in each training step to avoid the local minimum condition. For a training step, the input vector *i-th* can be represented as follows:

$$Z_{\text{in}}^{i} + Z_{\text{in}}^{0} + \Delta n_i, \tag{7}$$

where Δ stands for the interference level (small scale value) and n_i is the noise vector randomly sampled in the ith training. The final estimated HR-HS image utilized for prediction is the fixed noise vector $Z^* = G_\theta(Z_{\text{in}}^{0})$, which is created by feeding perturbed inputs into a neural network with coefficient G_θ.

This deep unsupervised fusion learning model employs noise vectors produced at random and sampled from a uniform distribution as input to provide low-level spatial statistics. But this research is less effective at identifying spectral and spatial correlations and is more challenging to optimize due to random noise vectors. We propose a solution to this issue in the next section. In the next part, we substitute observed

LR-HS and HR-RGB images for entirely artificial noise. Additionally, we approximate the degradation operation using two distinctive convolutional layers that can be applied as learning or fixed degradation models for a variety of real-world scenarios.

2.4.2 The fusion context

To deal with the mentioned problems, we improved the DUFL model above. The underlying prior structure of HR-HS images is reflected by an internally designed network structure in the deep self-supervised HS image reconstruction (DSSH) framework, which also learns the network parameters exclusively using observed LR-HS and HR-RGB images. In the proposed DSSH framework, we use the observed fusion context in network learning to gain insight into specific spatial and spectral priorities given the observed images: X reflecting hyperspectral properties of the underlying HR-HS image although with lower spatial-resolution, and Y showing the high-resolution spatial structure although with fewer spectral channels. To be more specific, we utilize an upconversion layer to first transform the LR-HS image to the same spatial dimension as the HR-RGB image before merging them, as seen below.

$$Z^0_{\text{in}} = Stack(UP(X), Y). \tag{8}$$

A simple fused context can be used as input, but this generally results in local minimum convergence. To train a more reliable model in this section that takes into account specific spatial and spectral priors, we add additional perturbations. The model is then represented as follows:

$$Z^i_{\text{in}} = Z^0_{\text{in}} + \lambda\mu, \tag{9}$$

where λ is a small number indicating the intensity of the perturbation and μ is a sample of a 3D tensor generated at random from a uniform distribution equal to the connection context Z^0_{in}. In this section, λ is set at 0.01 and reduced by half every 1000 steps throughout the training phase. The perturbation is applied to the generative network G_θ during each training phase.

Our suggested approach is capable of using any DCNN architecture for the G_θ generative network construction. Potential HR-HS images frequently have complicated spectra, expressive patterns, and rich textures, all of which demand the full modeling power of the generative network G_θ. Significant advancements have been achieved in generating higher natural images [28], and several generative architectures have been presented, for instance in adversarial learning situations [29].

2.5 Degradation modules

2.5.1 Non-blind degradation module

We apply degradation operations to get approximations of the LR-HS and HR-RGB images from the HR-HS images predicted by the generative network in order to provide evaluation criteria for training the network. However, this part of the network is removed and cannot be included in an integrated training system if only mathematical operations are utilized to approximate the degraded model. In this work, after constructing the backbone, we approximate the degradation model as a conventional learning system utilizing two parallel blocks. To specifically accept blurred and downsampling transformations, we modified the conventional deep convolutional layers. We apply the same kernel to various spectral bands in the depth-wise convolution layer and set the step space expansion coefficients and bias terms to "false" since the identical blurring and downsampling operations are applied to each spectral band in a real scene. The blurring and downsampling transformations' equations are written as follows:

$$\hat{X} = f_{SDW}(G_\theta(Z_{in})) = k_{SDW} \otimes G_\theta(Z_{in})^{(Spa)} \downarrow, \tag{10}$$

where the convolution layer's specific depth performs the role of $f_{SDW}(\bullet)$. To be more precise, we refer to the same kernel that was used in the depth-wise convolution layer to convolve $G_\theta(Z_{in})$ in the HR-HS images generated with each channel independently as $k_{SDW} \in R^{1 \times 1 \times s \times s}$. False bias is accomplished by using conventional two-dimensional mathematical convolution and nearest downsampling operations to transform the spatial expansion factor of $f_{SDW}(G_\theta(Z_{in}))$. If the spatially degraded blur kernel is known, we simply set the values to be trained as false values and initialize the weights of each layer based on the known kernel. Similar to this, we simply automatically learned kernel weights of 1*1 during the network training phase or assigned kernel weights of $f_{Spe}(\bullet)$ based on the known RGB camera CSF. Additionally, we employ a conventional convolution kernel with output channels of 3 and a kernel size of 1*1to implement the spectral transform. We similarly set the stride to 1 and the bias term to false, as shown in the following expression.

$$\hat{Y} = f_{Spe}(G_\theta(Z_{in})) = k_{Spe} \otimes G_\theta(Z_{in})^{(Spe)} \downarrow, \tag{11}$$

where the activity of the spectral convolution layer is indicated by $f_{Spe}(\bullet)$. The detailed spectra of the obtained HR-HS images are transformed into degenerate RGB images using the convolution kernel $k_{Spe} \in R^{L \times 3 \times 1 \times 1}$. Additionally, the kernel of k_{Spe} minimization that needs to be trained has the same dimension as the $C^{(Spec)}$ that represents the spectrum sensitivity function of an RGB sensor, allowing us to approximate it in our joint network. These two modules can be used concurrently in our integrated learning model by employing the mentioned framework.

2.5.2 (semi-) blind degradation module

This section focuses on automatically learning the transform parameters of the convolutional blocks embedded in the unknown decomposition. For spatially semi-blind, the weight parameter of k_{SDW} in Eq. (10) can either be automatically learned when the blur process is unknown while the weight parameter of k_{Spe} can be predetermined by changing to the parameter of a known CSF kernel. Thus, we can easily extract the approximation LR-HS image from the generated HR-HS image G_θ using a specified deep convolutional layer f_{SDW} with a fixed k_{Spe} convolutional kernel. Similarly, it is adaptable to implement the opposite operation to achieve a spectrally semi-blind process. Hence, these two modules can be learned concurrently in our integrated learning framework as a blind degradation module. As a result, the investigated learning model is extremely adaptable and simple to fit into many real-world scenarios. The loss function that was used to train our deep self-supervised network can be rebuilt as follows by substituting the decomposition operation with an improved convolutional block.

$$\left(\theta^*, \theta^*_{SDW}, \theta^*_{Spe}\right) = \underset{\theta}{\text{argmin}} \alpha\beta_1 \|X - f_{SDW}(G_\theta)(Z_{in}))\|_F^2 + (1-\alpha)\beta_2 \|Y - f_{Spe}(G_\theta)(Z_{in}))\|_F^2 \quad \text{s.t.} 0 \le G_\theta(_{Z_{in}} \le 1 \forall i. \tag{12}$$

As can be observed from Eq. (12), in order to rebuild the target well, we learn the generative network parameters rather than directly optimizing the underlying HR-HS image. In our network optimization procedure, the generative network G_θ is trained using only test image pairs (i.e., observed LR-HS and HR-RGB images), and no HR-HS images are provided. This can be seen as a "zero-shot" self-supervised learning method [30]. As a result, we refer to our model as a self-supervised learning model for HSI-SR.

3. Experiment results

3.1 Experimental settings

3.1.1 Datasets

The efficiency of the suggested method was evaluated using two benchmark HSI datasets, namely, CAVE [31] and Harvard [32]. 32 HS images with a spatial resolution of 512 × 512 are included in the CAVE dataset, which includes various real-world materials. The Harvard dataset includes 50 images of various natural settings, each with a resolution of 1392/1040 pixels and 31 bands of spectral-resolution between 420 and 720 nm. In the experiments, a part of the 1024 × 1024 sub-image in the top left corner of the Harvard dataset's original HS image was cropped, resulting in a 512 × 512 -pixel image that served as the HS image's main basis. Using different spatial extraction factors (8 and 16) for the bicubic degradation, the observed LR-HS images were generated from the actual HS images of the two datasets, yielding sizes of 64 × 64 × 31 and 32 × 32 × 31. The observed HR-RGB images were also generated by multiplying the HR-HS image by the spectral Nikon D700 camera response function [9].

3.1.2 Evaluation metrics

The proposed method is evaluated against various state-of-the-art methods using five widely used metrics, including root-mean-square error (RMSE), signal-to-noise ratio (PSNR), structural similarity index (SSIM), spectral angle mapping (SAM), and relative dimensional global error (ERGAS). The generated HR-HS image and the ground-truth image were both acquired from the same spatial position. The recovered HR is measured by RMSE, PSNR, and ERGAS which are quantitatively distinct from the reference image to assess the spatial accuracy. Then, SAM offers the average spectral angle of the two spectral vectors to show the spectral accuracy. Additionally, SSIM was employed to evaluate how much the spatial organization of the two images resembled one another. A greater PSNR or SSIM and a lower RMSE, ERGAR, or SAM often indicate superior performance. Bold values mean promising results.

3.1.3 Details of the network implementation

Pytorch has adopted the suggested approach. The input noise was first set to the same size as the HR-HS image that would be generated. Utilizing the Adams

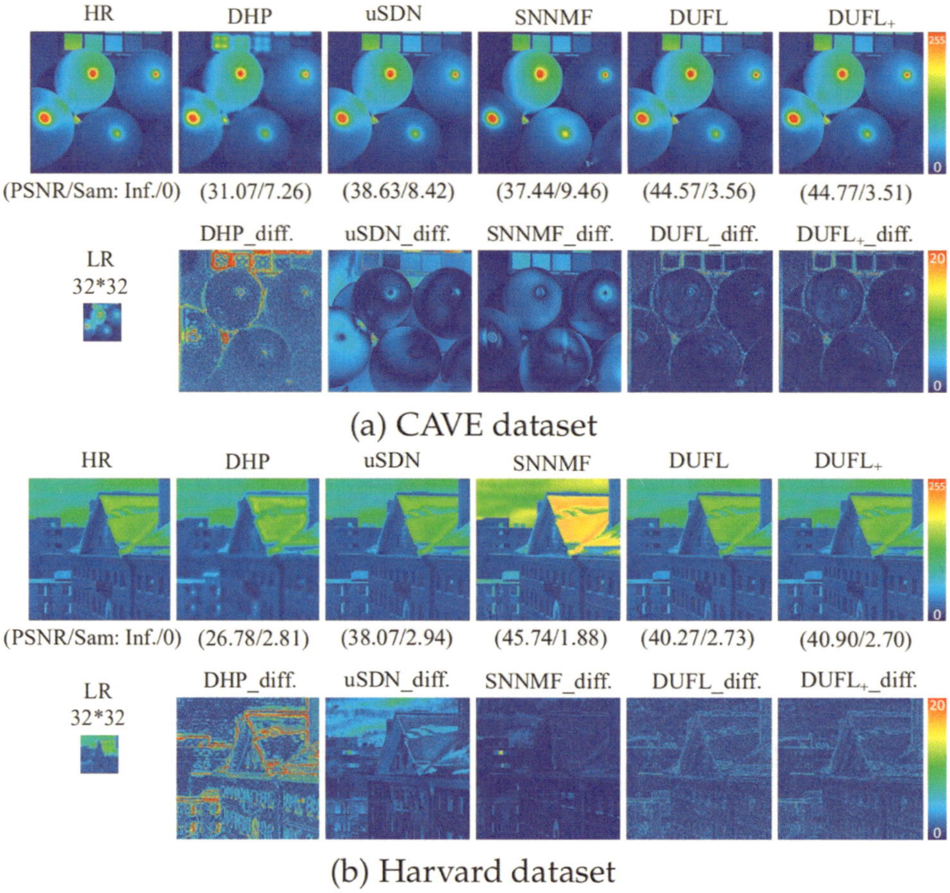

Figure 2.
Visualization of the DHP [36], uSDN [20], SNNMF [37], and difference images between the proposed DUFL+ method and the ground-truth/reconstructed images in CAVE and Harvard datasets for an up-scale factor 16.

10 test images from the Harvard dataset were compared because supervised deep learning methods need training examples to learn the model. The results of the comparison between the CAVE and Harvard datasets are shown in **Table 2**, with two spatial expansion factors: 8 and 16. It is clear from **Table 2** that our proposed method can perform noticeably better than unsupervised methods based on deep learning, as well as better than supervised methods. And the visualization results are shown in **Figure 3**.

3.2.3 Comparison with (semi-)blind methods

Our proposed method is exploited in a unified framework, which is capable of reconstructing the HR-HS image from the observations not only with the known spatial and spectral degradation operations but also with the unknown spatial or spectral degradation operations or both unknown. Thus, our proposed method can be implemented in a semi-blind setting (the unknown spatial downsampling kernel for

Up-scale factor = 8											
Data		CAVE					Harvard				
Method		RMSE ↓	PSNR ↑	SSIM ↓	SAM ↓	ERGAS↓	RMSE ↓	PSNR ↑	SSIM ↓	SAM ↓	ERGAS ↓
Supervised	SSFNet [13]	1.89	44.41	0.991	3.31	0.89	2.18	41.93	0.991	4.38	0.98
	ResNet [14]	1.47	45.90	0.993	2.82	0.79	1.65	44.71	0.984	2.21	1.09
	DHSIS [16]	1.46	45.59	0.990	3.91	0.73	1.37	46.02	0.981	3.54	1.17
Unsupervised	uSDN [20]	4.37	35.99	0.914	5.39	0.66	2.42	42.11	0.987	3.88	1.08
	DHP [36]	7.60	31.40	0.871	8.25	4.20	7.94	30.86	0.803	3.53	3.15
	GDD [22]	1.68	44.22	0.987	3.81	0.96	1.30	47.02	0.990	1.94	0.90
	DUFL (Our)	2.10	42.53	0.978	5.30	1.12	2.15	42.63	0.975	2.32	1.01
	DUFL+ (Our)	2.09	42.39	0.977	4.54	0.91	2.75	40.41	0.965	0.03	0.58
	DSSH (Our)	1.44	45.61	0.992	3.27	0.79	1.17	48.27	0.993	1.75	0.77
Up-scale factor = 16											
Supervised	SSFNet [13]	2.18	41.93	0.991	4.38	0.98	1.94	43.56	0.980	3.14	0.98
	ResNet [14]	1.93	43.57	0.991	3.58	0.51	1.83	44.05	0.984	2.37	0.59
	DHSIS [16]	2.36	41.63	0.987	4.30	0.49	1.87	43.49	0.983	2.88	0.54
Unsupervised	uSDN [20]	3.60	37.08	0.969	6.19	0.41	9.31	39.89	0.931	4.65	1.72
	DHP [36]	11.31	27.76	0.805	10.66	3.09	10.38	38.44	0.754	4.57	2.08
	GDD [22]	2.12	42.24	0.983	4.41	0.61	1.66	44.64	0.986	2.50	0.64
	DUFL (Our)	2.60	40.75	0.970	6.42	0.70	9.46	38.14	0.876	8.52	7.71
	DUFL+ (Our)	2.95	40.56	0.948	2.25	1.15	3.12	39.79	0.945	2.76	0.66
	DSSH (Our)	1.76	43.84	0.999	3.76	0.49	1.32	47.16	0.992	1.99	0.47

Table 2.
Compared results of the deep non-blind learning-based methods with DUFL and DSSH methods in the CAVE and Harvard datasets for up-scale factors: 8 and 16.

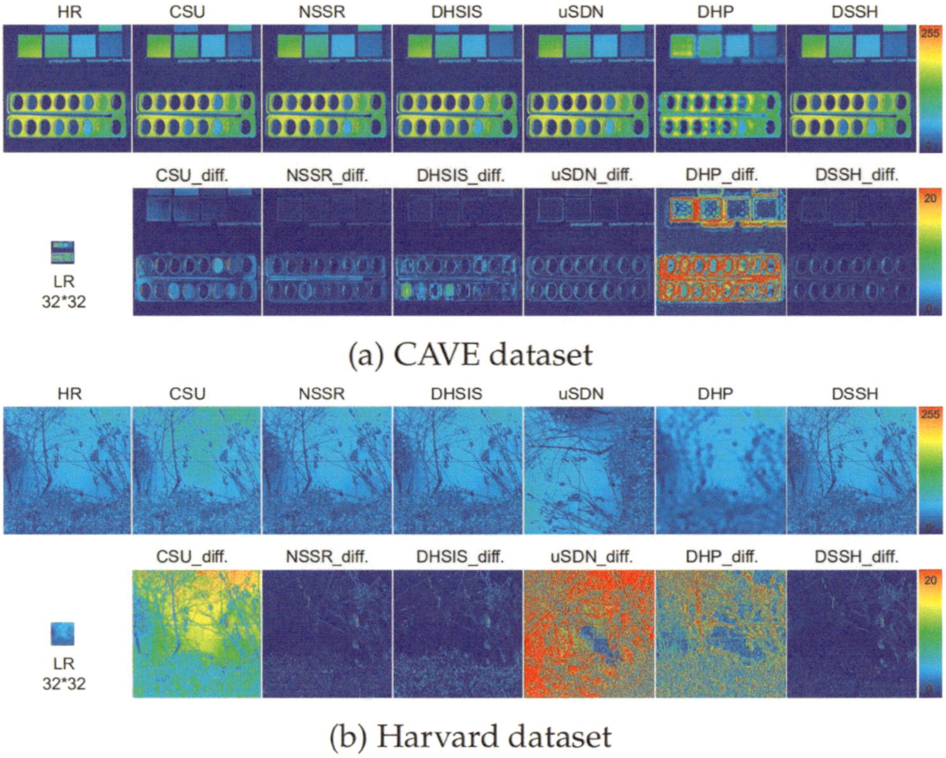

Figure 3.
Visualization of the traditional optimization-based method: CSU [9] and NSSR [7], the supervised deep learning-based methods: DHSIS [16], and the unsupervised deep learning-based methods: uSDN [20], DHP [36], and the proposed DSSH method in the CAVE and Harvard datasets for an up-scale factor 16.

LR-HS image or the unknown CSF for HR-RGB image). Consequently, our suggested solution can also be used in total blind mode (unknown spatial degradation operations for LR-HS images and unknown CSF for HR-RGB images). The compared results using our proposed method with semi-blind and complete-blind settings, the state-of-the-art unsupervised semi-blind methods: UAL method [23] for spatial blind only, and the spatial blind implementation of NSSR [7] via setting the incorrect spatial kernel, have been given in **Table 3**.

3.2.4 Ablation study

We adjusted the hyperparameters α to 0.3, 0.5, and 0.7 in order to assess the impact of various data circumstances on the loss function of the DUFL method. The comparative results are shown in **Table 4**. The quantitative measurements of our DUFL+ method, PSNR, SAM, and ERGAS, are also shown in **Table 4**, and they demonstrate that the performance of overfitting is not significantly affected by the specific assignment of the hyperparameter α. Similarly, the performance of the DSSH reconstruction method in the ablation study was then evaluated by adjusting α between 0 and 1.0 with an interval of 0.2, and the compared results are shown in **Table 5**.

Method	Real downsampling kernel	CAVE					Harvard				
		RMSE ↓	PSNR ↑	SSIM ↓	SAM ↓	ERGAS ↓	RMSE ↓	PSNR ↑	SSIM ↓	SAM ↓	ERGAS ↓
NSSR (Bic) [7]	Bicubic	3.41	38.03	0.968	5.35	1.52	2.76	39.77	0.981	2.00	1.30
NSSR (Ave) [7]	Average	2.76	39.77	0.981	2.00	1.30	3.27	38.55	0.972	5.17	1.78
UAL [23]	K1	1.85	43.23	0.986	6.72	—	2.08	42.38	0.982	2.67	—
	K2	2.01	42.72	0.986	6.78	—	—	—	—	—	—
DSSH (Our) (Spatial blind)	K1	1.47	45.14	0.990	3.54	0.66	1.15	47.59	0.994	1.70	0.78
	K2	1.56	44.71	0.989	3.64	0.69	1.12	47.75	0.994	1.70	0.79
	Bicubic	1.70	44.05	0.988	3.70	0.75	1.33	46.28	0.992	1.95	0.93
DSSH (Our) (Spectral blind)	Bicubic	1.64	44.36	0.989	3.66	0.72	1.28	46.67	0.992	1.86	0.89
DSSH (Our) (Complete blind)	Bicubic	1.68	44.10	0.988	3.72	0.74	1.32	46.44	0.992	1.91	0.91

Table 3.
Compared results of the (semi-)blind methods with DUFL and DSSH methods in the CAVE and Harvard datasets for an up-scale factor 8.

Up-scale Factor	α	CAVE			Harvard		
		PSNR ↑	SAM ↓	ERGAS ↓	PSNR ↑	SAM ↓	ERGAS ↓
8	0.3	42.19	5.09	0.95	43.07	2.16	0.93
	0.5	42.91	4.40	0.86	41.68	2.19	1.06
	0.7	42.16	4.75	0.92	41.85	2.18	1.09
16	0.3	40.74	5.71	0.55	40.95	2.90	0.66
	0.5	40.75	5.87	0.54	40.79	2.70	0.62
	0.7	40.42	5.64	0.58	41.90	2.48	0.52

Table 4.
Ablation results of the DUFL+ method with different weights α values of 0.3, 0.5, and 0.7 in the CAVE and Harvard datasets for up-scale factors: 8 and 16.

Dataset	CAVE				
α	RMSE ↓	PSNR ↑	SSIM ↓	SAM ↓	ERGAS ↓
0.0	25.98	19.97	0.631	40.02	12.50
0.2	1.52	44.99	0.990	3.24	0.67
0.4	1.45	45.45	0.991	3.16	0.63
0.5	1.46	45.35	0.991	3.13	0.64
0.6	1.49	42.26	0.991	3.15	0.66
0.8	1.47	45.20	0.991	3.13	0.66
1.0	3.33	38.36	0.961	4.73	1.51

Table 5.
Ablation results of the DSSH method with different weights α values of 0.0 to 1.0 in the CAVE and Harvard datasets for an up-scale factor 8.

4. Conclusions

In order to address the super-resolution issue for hyperspectral images, we provide an unsupervised deep hyperspectral image super-resolution framework. A deep convolutional neural network is used to automatically learn the spatial and spectral features of latent HR-HS images from perturbed noisy input data and the fusion context that naturally collects a significant quantity of low-level image statistics. A special depth-wise convolution layer is designed to achieve degenerate transformations between observations and desired targets, and this generates a universally learnable module that only uses low-quality observations. Without requiring training samples, the proposed unsupervised deep learning framework can efficiently take advantage of the HR spatial structure of HR-RGB images and the detailed spectral characteristics of LR-HS images to deliver more accurate HS image reconstruction. We simply train the network parameters using the observed LR-HS and HR-RGB images and a generative network structure to reconstruct the underlying HR-HS images. Extensive research using the CAVE and Harvard datasets demonstrate promising results in the quantitative evaluation.

Author details

Zhe Liu and Xian-Hua Han*
Graduate School of Sciences and Technology for Innovation, Yamaguchi University, Yamaguchi, Japan

*Address all correspondence to: hanxhua@yamaguchi-u.ac.jp

By adding these two regularization terms with two convex subproblems helps to upgrade the performance of the existing CNMF method. However, sometimes performance degradation may occur in the CO-CNMF algorithm as the noise level increases. Therefore, it is necessary to add image denoising and spatial smoothing constraints with this fusion method.

Yang *et al.* in 2019 [17] introduced a total variation and signature-based (TVSR) regularizations CNMF method named as TVSR-CNMF. The TV regularizer is added to the abundance matrix to ensure the images spatial smoothness. Similarly, a signature-based regularizer (SR) is added with the endmember matrix for extracting high-quality spectral data. So, this method helps to reconstruct a hyperspectral image with good quality in spatial and spectral data.

Yang *et al.* in 2019 [18] introduced a sparsity and proximal minimum-volume regularized CNMF method named as SPR-CNMF. The minimum-volume regularizer controls and minimizes the distance between selected endmembers and the center of mass of the selected region in the image to reduce the computational complexity. It redefines the fusion method at each iteration until reaches the simplex with minimum volume. This method improves the fusion performance by controlling the loss of cubic structural information.

After being influenced by this work, we implemented an unmixing-based fusion algorithm named fully constrained CNMF (FC-CNMF). This method is a modified version of CNMF by including all spatial and spectral constraints available in the literature. In our method, a simplex with minimum volume constraint is imposed with the endmember matrix to exploit the spectral information fully. Similarly, sparsity and total variation constraints are incorporated with the abundance matrix to provide dimensionality reduction and spatial smoothness to the image. Finally, we evaluated the quality of the fused image obtained by FC-CNMF against the methods discussed in the literature using some standard quality measures. From these evaluations, we understood that our method shows better performance by yielding high-fidelity in the reconstructed images.

These traditional approaches reconstruct the high-resolution hyperspectral image by fusing the high-quality data from hyperspectral and multispectral images. However, to improve the quality of the reconstructed images, these approaches use different constraints such as sparsity, minimum volume simplex, and total variance regularization, etc. The performance and quality of the reconstructed HS image are highly influenced by these constraints and therefore our existing methods still have an ample space to enhance the quality of HSI.

2.2 Deep learning methods

Deep learning (DL) is a subbranch in machine learning (ML) and has shown remarkable performances in the research field, especially in the area of image processing and computer vision recently. DL is based on an artificial neural network that has been widely used in different areas such as super-resolution, classification, image fusion, object detection, etc. DL-based image fusion methods have the ability to extract deep features automatically from the image. Therefore, DL-based methods overcome the difficulties that are faced during the conventional image fusions methods and make the whole fusion process as easier and simple.

A deep learning-based HS-MS image fusion concept was first introduced by Palsson *et. al* in 2017 [19]. In this method, they used a 3-D convolutional neural network (3D-CNN) to fuse LR–HS and HR–MS image to construct HR-HS image.

This method improves the quality of hyperspectral image by reducing noise and the computational cost. In this paper, they focused on enhancing the spatial data of LR–HS image without any changes in the spectral information and it caused the degradation of spectral data [19].

Later, Masi *et al.* in 2017 [20] proposed a CNN-architecture for image super-resolution, which uses deep CNN for extracting both spatial and spectral features. Deep CNN is used to acquire features from HSI with a very complex spatial-spectral structure. But in this paper, authors used deep CNN with single branch CNN architecture which is difficult to extract the discriminating features from the image.

To overcome this drawback, Shao and Cai in 2018 [21] designed a fusion method by extending CNN with depth of 3D-CNN for obtaining better performance while fusion. For implementing this, they used a remote sensing image fusion neural network (RSIFNN) that uses two CNN branches separately. One branch extract the spectral and the other extract the spatial data from the image. In this way, this method helps to exploit the spectral as well as spatial information from the input images to reconstruct high spectral and spatial resolution hyperspectral image.

Yang *et.al* in 2019 [22] introduced a deep two-branch CNN for HS–MS fusion. This method uses a two-branch CNN architecture for extracting spectral and spatial features from LR–HSI and HR–MSI. These extracted features from two branches of CNN are concatenated and then passed to the fully connected convolution layer to obtain HR–HSI. In all the conventional fusion methods, HR–HSI is reconstructed in a band-by-band fashion whereas in CNN concepts all bands are reconstructed jointly. Therefore, it helps to reduce the spectral distortion that occurs in the fused image. But this method uses fully connected layer for image reconstruction that is heavily weighted layer and it increases the network parameters.

Chen *et al* in 2020 [23], introduced a spectral–spatial features extraction fusion-CNN (S2FEF- CNN) which extracts joint spectral and spatial features by using three - S2FEF blocks. The S2FEF method use 1D and 2D convolution network to extract spectral and spatial features and fuse these spectral and spatial features. This method uses fully connected network layer for dimensionality reduction, and it further reduces the network parameters during the fusion. This method shows good results with less computational complexity compared to all other deep learning-based fusion method.

Although the deep learning-based fusion methods achieved tremendous improvement in their implementation, however, all these methods still possess many drawbacks [24]. As the network goes deeper, its performance gets saturated and then rapidly degrades. This is because, in DL method, each convolution layer takes inputs from the output of the previous layers, so when it reaches the last layer, a lot of meaningful information obtained from the initial layers will be lost. The information loss tends to get worse when the network is going deeper in architecture. This will bring some negative effects such as overfitting of data and this effect is called vanishing gradient problem [25].

Due to the vanishing gradient problem, the existing deep learning-based fusion could not be able to extract the detailed features from high dimensional images. He *et al* in ref., [26], introduced a deep network with residual learning to address the vanishing gradient problem. In this framework, a residual block is added between the layers to diminish the performance degradation. The networks with these concepts are called residual networks or ResNets. Therefore, in this work, our aim is to invoke this ResNet architecture into the standard CNN to exploit more detailed features from both spatial and spectral data of HSI.

3. Materials and methods

3.1 Dataset

The four real datasets such as Washington DC mall, Botswana, Pavia University, and Indian Pines are used in this work. The Washington DC Mall dataset is a well-known dataset captured by HYDICE sensor, which acquired a spectral range from 400 to 2500 nm have 1278×307 pixel size and 191 bands. The Botswana dataset which is captured by Hyperion sensor acquired over the Okavango delta in Botswana, which acquired a spectral range from 400 to 2500 nm with 1476 × 256 pixel size and 145 bands. The Pavia University dataset was captured by the reflective optics spectrographic imaging system (ROSIS-3) at the University of Pavia, northern Italy, in 2003. It has a spectral range from 430 to 838 nm and has a 610 × 340 pixel size and 103 bands. Finally, the dataset AVIRIS Indian Pines was captured by AVIRIS sensor over the Indian Pines test site in northwestern Indiana, USA, in 1992. It acquired a spectral range from 4 to 2500 μm having 512 × 614 pixel size and 192 bands [26]. All these datasets have been widely used in earlier spectral unmixing-based fusion research.

3.2 Convolution neural networks

Convolutional neural networks (CNN) have an important role in deep learning models. CNN specially built an algorithm that is designed to work with images to extract deep features from the image through convolution. The convolution is a process that applies a kernel filter across every element of an image to understand and react to each element within the images. This concept of convolution is more helpful during the extraction of specific features from high dimensional images. A convolutional network architecture is composed of an input layer, an output layer, and one or more hidden layers. The hidden layers are combination of convolution layers, pooling layers, activation layers, and normalization layers. These layers automatically detect essential features without any human supervision. So it is considered as a powerful tool for image processing [27].

A. Convolution layer

The convolution layer is used to extract various features from the input image with the help of filters. In convolution layer, mathematical operation is performed between the input image and the filter with m × m kernel size. This filter is sliding across the input image to calculate the dot product of the filter and part of the image. This process is repeated for convolving the kernel to all over the image and the output of the convolution operation is called a feature map. This feature map includes all essential information about the image such as the boundary and edges of objects etc. [28].

B. Pooling layer

The convolution layer is followed by a pooling layer, which reduces the size of the feature map by maintaining all the essential features. There are two types of pooling layers such as max pooling and average pooling. In Max pooling, the largest element is taken from the feature map whereas in the average pooling calculates the average of the element in the feature map [28].

C. Activation function

One of the most important characteristics of any CNN is its activation function. There are several activation functions such as sigmoid, tanH, softmax, and ReLU, and all these functions have their own importance. The ReLU is the most commonly used activation function in DL that accounts for the nonlinear nature of the input data [28].

3.3 Residual network (ResNet)

A residual network is formed by stacking several residual blocks together. Each residual block consists of convolution layers, batch normalization, and activation layers. The batch normalization process the data and brings numerical stability by using some scaling techniques without distorting the structure of the data. The activation layer is added into the residual network to help the neural network to learn more complex data. The CNN or deep learning method uses ReLU (rectified linear unit) function in the activation layer to accommodate the nonlinearity nature of the image data while providing the output. The residual blocks allow to flow information from the first layer to the last layers of the network by adding residual or skip connection strategy. Therefore, ResNet can effectively utilize features of the input data to the output of the network and thus alleviate gradient vanishing problems.

Let x be the input to the residual block, after processing the information x with two-stacked convolution layers of a residual unit, obtains $F(W_1x)$, where W is the weight given to the convolution layer. In ResNet, before giving an output of one convolution layer $F(W_1x)$ as input of the next layer by adding the x term, which is the input parameters of previous residual block, to provide an additional identity mapping information called as skip connection. Therefore the general formulation of a residual block can be represented as follows:

$$y = F(W_i\, x) + x \tag{2}$$

Here x is an input and y is the output of the residual unit. Then y is a guaranteed input to the next residual block. The function $F(W_i\, x)$ represents the output of each convolution layer, and W_i is the weight associated with i^{th} residual blocks. **Figure 1** uses two convolution layers for the residual unit, so the output from this residual layer can be written as:

$$F(xW) = W_2ReLU(W_1x) \tag{3}$$

Where $ReLU$ represents the nonlinear activation function rectified linear unit (ReLU), W_1 and W_2 are the weight associated with convolution layers 1 and 2 of the residual block A. Deep residual networks consist of many stacked residual blocks and each block can be formulated in general as follows:

$$x_{i+1} = F(x_l W_l) + x_i \tag{4}$$

Where F is the output from residual block with l stacked convolution layer and x_i is the residual connection to the ith residual block, then x_{i+1} become the output of the ith residual block, which is calculated by a skip connection and element-wise

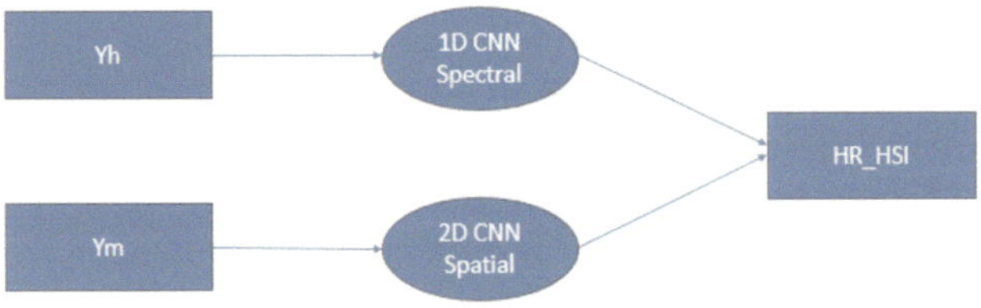

Figure 1.
HS–MS fusion using CNN.

multiplication. After passing through the ReLU activation layer, the output residual network can be represented as:

$$y = ReLU(x_{i+1}) \tag{5}$$

4. Problem formulation

A high-resolution hyperspectral image $Z \in \mathbb{R}^{L \times N}$ with L spectral band and N pixels. The observed LR–HSI is obtained by downsampling the spatial quality of Z with a gaussian blur factor d is represented as $Y_h \in \mathbb{R}^{L \times N_{/d}}$ with L bands and $N_{/d}$ pixels. Similarly, the observed HR–MSI is obtained by downsampling the spectral quality of Z and it is represented as $Y_m \in \mathbb{R}^{L_m \times N}$ with L_m bands and N pixels, where $L_m < L$ [27]. Therefore, the hyperspectral image can be mathematically modeled as:

$$Z = EA + R \tag{6}$$

Where, Z is the original referenced images, E and A are the endmember, abundance matrices, and R is the residual matrix respectively.

The observed Y_h and Y_m are spectrally and spatially degraded versions of image Z is further mathematically represented by:

$$Y_m \approx SZ + R_m \tag{7}$$

$$Y_h \approx ZB + R_h \tag{8}$$

Where $B \in \mathbb{R}^{N \times N_{/d}}$ is a Gaussian blur filter with blurring factor d used to blur the spatial quality of the referenced hyperspectral image Z to obtain LR–HSI, Y_h. The spectral response function, $S \in \mathbb{R}^{L_m \times L}$ is used to downsampling the spectral quality of the referenced hyperspectral image Z to obtain HR–MSI, Y_m. The term $L_m$means the number of spectral bands used in the multispectral image after downsampling. In this work, referenced image Z is downsampled by its spectral values using standard L and sat 7 multispectral image that contains a high-quality visual image of Earth's surface as HR–MSI with $L_m = 7$ [28]. Both B and S are spared matrices containing zeros and ones. In general, the residual matrix R_m and R_h are assumed as zero-mean Gaussian noises in the literature, Therefore, the original CNMF method is shown as:

$$\text{CNMF}\ (E, A) = \|Y_h - (EA_h)\|_F^2 + \|Y_m - (E_m A)\|_F^2 \tag{9}$$

However, in this work, we make use of the residual term R_m and R_h as a nonnegative residual matrix to account for the nonlinearity effects in the image fusion [29].

Since the objective function for the original CNMF method expressed in the Eq. (9) can be re-written as:

$$\text{CNMF (E, A, R)} = \|Y_h - (EA_h + R_h)\|_F^2 + \|Y_m - (E_m A + R_m)\|_F^2 \tag{10}$$

Therefore the Eq. (10) represents the proposed model of the HS–MS fusion by including the nonlinearity nature of the image. To implement this model, we use standard deep neural network architecture CNN and ResNet. For further enhancement of the proposed method, we implemented modified architecture of ResNet with different stacked layers and multiple skip connections.

5. Problem implementation

5.1 CNN fusion architecture

In CNN architecture, 1D CNN convolution operation is performed over the observed HS image Y_h of dimension $L_h x N_h$ with L_h spectral band and N_h number of pixels in the image with the help of filter to obtain the spectral data. In the same way, 2D CNN convolution operation is performed over the observed MS image is denoted by Y_m of dimension L_m x N_m, with L_m spectral bands and N_m number of pixels in the image to obtain the spatial data. Finally, the high spectral component obtained from Yh and high spatial component obtained from Ym are fused together to reconstruct a high HR-HSI. The entire deep neural network-based HS–MS fusion is shown in **Figure 1**.

In CNN architecture, the *Conv1D*() convolution filter with kernel size r having weight v are used for extracting spectral data from LR–HSI, Y_h are represented as follows:

$$f_{spec} = Conv1D(ReLU(\ F(v_i\ Y_h))) \tag{11}$$

Similarly, the *Conv2D()* convolution filter with kernel size r × r having weight w are used for extracting spatial data from HR–MSI, Y_m image are represented as:

$$f_{spat} = Conv2D\big(ReLU\big(\ F\big(w_{ij}\ Y_m\big)\big)\big) \tag{12}$$

The two convolutional layers use ReLU (rectified linear unit) activation functions, i.e., *ReLU* (x) = max(x, 0), to provide nonlinear mapping of data. Finally, fuse the extracted spatial and spectral features to get high-quality reconstructed image as shown in Eq. (4).

$$\mathrm{F} = ReLU\Big(f_{spec} \times f_{spat}\Big) \tag{13}$$

To implement this CNN fusion architecture, we use two convolution networks such as 1D and 2D convolution. Both 1D and 2D convolution uses the same number of convolution layers and kernel size. Each network uses four convolution layers with 32, 64, 128, and 256 filters. The kernel size of 3 × 3 and 1 × 3 are used for 2D CNN and 1D CNN for extracting spatial and spectral information about the image. Therefore, the architecture and parameters of CNN HS-MS fusion are shown in **Table 1**.

Layer		Filter	Kernel size	Stride	Padding	Activation
Conv 1	Conv 1D	32	1 × 3	1	Same	ReLU
	Conv 2D	32	3 × 3			
Conv 2	Conv 1D	64	1 × 3	1	Same	ReLU
	Conv 2D	64	3 × 3			
Conv 3	Conv 1D	128	1 × 3	1	Same	ReLU
	Conv 2D	128	3 × 3			
Conv 4	Conv 1D	256	1 × 3	1	Same	ReLU
	Conv 2D	256	3 × 3			
Output layer	Conv 1D	1	1 × 1	1	Same	ReLU
	Conv 2D	1	1 × 1			

Table 1.
The Simple CNN Fusion Architecture.

In CNN, each layer takes its input as the output from the previous layer and it introduces lose information as the network architecture goes in deeper. This problem in deep neural network leads to overfitting of data, and it is known as vanishing gradient problem [24]. To overcome this, we implemented HS-MS fusion using an alternative ResNet-based network architecture. In ResNet, we introduced the skip connection between two convolution layers. This skip connection helps to map the identity of information throughout the deep convolution network.

5.2 Resnet fusion architecture

The ResNet fusion architecture for HS–MS fusion uses residual or skip connection which helps to improve the feature extraction capability from the images. For implementation, we use 1D ResNet to extract the spectral features from the LR–HSI and 2D ResNet for extracting spatial features from HR–MSI. Both 1D and 2D ResNet architecture consists of three residual blocks each having two convolutional layers and 64 filters as shown in **Figure 2**. A3 × 3 kernel size for 2D Resnet and 1 × 3 kernel size for 1D Resnet are used for extracting the spatial and spectral data from MSI and HSI. Each residual block has ReLU activation layer to accommodate the nonlinearity constraints included in the proposed hyperspectral image fusion model as explained in Eq. (10). Finally, the feature embedding and image reconstruction process are performed using another 2D CNN.

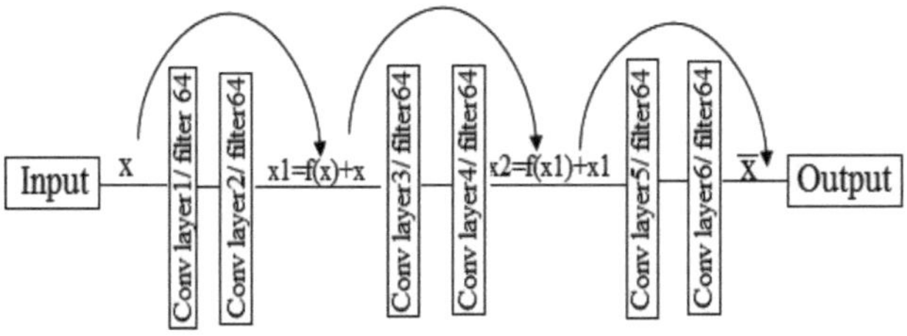

Figure 2.
Residual block with two stacked layer.

A. Spectral generative network

The spectral data from hyperspectral image Y_h is extracted using 1D ResNet connection. Initially, spectral data are extracted from LR–HSI using 1D CNN and then mapping the residual connection $r(Y_h)$ with the stacked convolution layers. Finally, the output from ID CNN and $r(Y_h)$ are given to the input of the next residual block and this process is repeated for an entire residual block in the ResNet. The entire process in 1D ResNet is shown mathematically as:

$$f\left(Y_{h_l}\right) = ReLU\left(W_l Y_{h_l}\right) \tag{14}$$

$$f_{spec}\left(Y_{h_l}\right) = f\left(Y_{h_l}\right) + r\left(Y_{h_l}\right) \tag{15}$$

Therefore, output of ith residual block is represented as:

$$f^{i}_{spec} = f^{i-1}_{spec}\left(Y_{h_l}\right) + r^{i-1}\left(Y_{h_l}\right) \tag{16}$$

Where, Y_h denotes the input LR- HSI data, i is the number of residual units i = 1,2,3 ... ..I and l are the number of convolution layer l = 1,2,3 ... ..l. The weight of convolution kernel is represented as W. Finally, *ReLU* an activation functions are exploited to introduce nonlinearities in the output of deep network as follows:

$$F_{spec} = ReLU\left(f_{spec}\right) \tag{17}$$

B. Spatial generative network

The spatial data from HR–MSI, Y_m is extracted using 2D ResNet. Initially, spatial data are extracted from HR–MSI using 2D CNN and then mapping the residual connection $r(Y_m)$ with the stacked convolution layers. Finally, the output from 2D CNN to $r(Y_m)$ is given to the input of the next residual block and this process is repeated for an entire residual block in the ResNet. The entire process in 2D ResNet is shown mathematically as:

$$f\left(Y_{m_l}\right) = ReLU\left(W_l Y_{m_l}\right) \tag{18}$$

$$f_{spat}\left(Y_{m_l}\right) = f\left(Y_{m_l}\right) + r\left(Y_{m_l}\right) \tag{19}$$

Therefore, output of the ith residual block is represented as:

$$f^{i}_{spat} = f^{i-1}_{spat}\left(Y_{m_l}\right) + r^{i-1}\left(Y_{m_l}\right) \tag{20}$$

Where, Y_m denotes the input HR- MSI data, i is the number of residual blocks i = 1,2,3 ... ..I and l are the number of convolution layer l = 1,2,3 ... ..L. The weight of the convolution kernel is represented as W. Finally, similar to spectral extraction *ReLU* is exploited to introduce nonlinearities in the spatial output of a deep network as follows:

$$F_{spat} = ReLU\left(f_{spat}\right) \tag{21}$$

C. Fusion of spectral-spatial data

The spectral data from LR–HSI and spatial data from HR–MSI are extracted using ResNet with size as (1x1x Spec) and (Spat x Spat x 1). After obtaining the spatial and spectral features, next step is to fuse this information by element-wise multiplication.

$$F_Z = F_{spec} \, x \, F_{spat} \tag{22}$$

Then, the feature embedding and image reconstruction are performed by using ReLU activation layer. The proposed ResNet Fusion framework is shown in **Figure 3**. Therefore, the final generated HR-HSI, Z can be written as:

$$Z = ReLU(F_Z) \tag{23}$$

D. Different stacked layers and skip connection

We also proposed an extension to the ResNet fusion architecture by varying the number of stacked convolution layers (2 to 4) in the residual block to increase the performance of the fusion using deep network. The 2-layer residual block contains two stacked convolution l ayer followed by ReLU activation layer. Similarly, in three-layer and four-layer residual blocks contain three and four-stacked convolution layers followed by ReLU activation layer. In addition to this, we utilize the ResNet fusion architecture by including different skip connections. The skip connection helps us to regulate the flow of information to a deeper network more effectively. For this, we use long skip and dense skip connections as shown in **Figure 4**. The long skip connections are designed by creating a connection between alternate residual layer i^{th} and $(i + 2)^{th}$ along with a short skip connection between every layer in the ResNet. In dense skip connection, each layer i obtain an additional input from all the preceding layers. Then, the layer i pass its own feature maps to all the subsequent layers. Using the

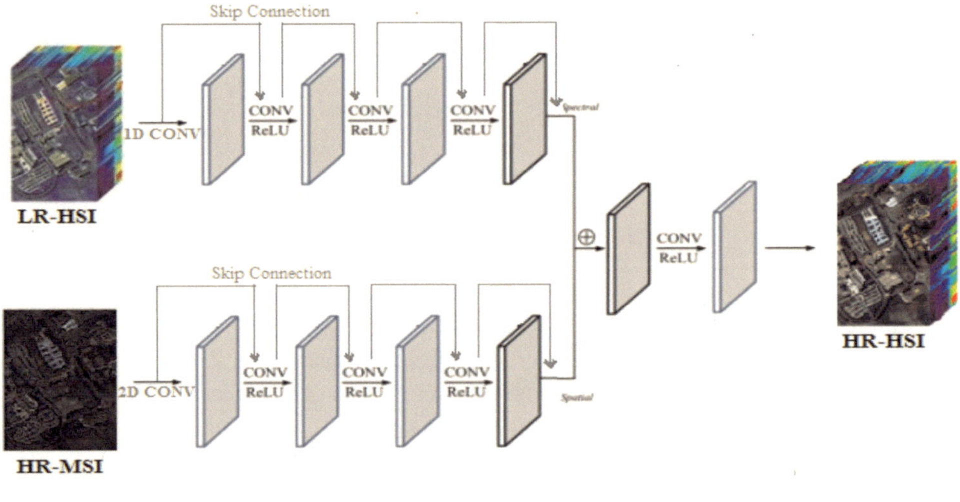

Figure 3.
The framework of the proposed ResNet Fusion architecture.

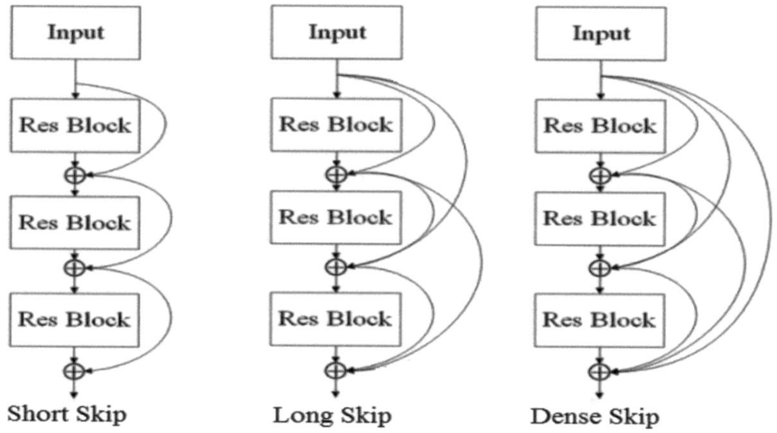

Figure 4.
Representation of short, long, and dense skip connection on ResNet.

dense skip connection, each layer in the ResNet receives feature maps from all the preceding layers and that limits the number of filters and network parameters for extracting deep features. In order to obtain high fidelity reconstructed image, we proposed a modified version of ResNet with long and dense skip connections shown in **Figure 4**.

In the **Figure 4** show three Resnet architecture, having three- residual blocks (Res Block), with three different types of skip connections. Algorithm 1 summarizes the procedures of our proposed ResNet fusion method.

Algorithm 1: Resnet Fusion

Input: LR-Hyperspectral image Y_h and HR-Multispectral image Y_m
begin
1. Extract spectral features from Y_h and spatial features from Y_m using ResNet
2. $r(Y_h) \leftarrow Y_h$ and $r(Y_m) \leftarrow Y_m$
3. **For each residual block in ResNet** $i = 1,2,3 \ldots .I$ # for each residual block
4. **for** each convolution layer l in the residual block $l = 2,3,4$ # for stacked convolution layer

   $\mathbf{f}(\mathbf{Y}_{\mathbf{h}_l}) = \boldsymbol{ReLU}(\mathbf{W}_l\mathbf{Y}_{\mathbf{h}_l})$
   $\mathbf{f}(\mathbf{Y}_{\mathbf{m}_l}) = \boldsymbol{ReLU}(\mathbf{W}_l\mathbf{Y}_{\mathbf{m}_l})$
   end for
   # add the residual connection
   $\mathbf{f}_{\mathbf{spec}}(\mathbf{Y}_{\mathbf{h}_l}) = \mathbf{f}(\mathbf{Y}_{\mathbf{h}_l}) + \mathbf{r}(\mathbf{Y}_{\mathbf{h}_l})$
   $\mathbf{f}_{\mathbf{spat}}(\mathbf{Y}_{\mathbf{m}_l}) = \mathbf{f}(\mathbf{Y}_{\mathbf{m}_l}) + \mathbf{r}(\mathbf{Y}_{\mathbf{m}_l})$
   $r(Y_h) \leftarrow \mathbf{f}_{\mathbf{spec}}(\mathbf{Y}_{\mathbf{h}_l})$
   $r(Y_m) \leftarrow \mathbf{f}_{\mathbf{spat}}(\mathbf{Y}_{\mathbf{m}_l})$
   end for
5. The extracted spectral features $\mathbf{F}_{\mathbf{spec}}$ of size (1x1x Spec) and spatial features $\mathbf{F}_{\mathbf{spat}}$ of size (Spat x Spat x1) are fused together by element-wise multiplication.
6. $\mathbf{F}_Z = \mathbf{F}_{\mathbf{spec}} \mathbf{\ x\ } \mathbf{F}_{\mathbf{spat}}$
7. Finally, generated HR-HSI after feature embedding and image reconstruction using relu activation layer.
8. $Z = \boldsymbol{ReLU}(\boldsymbol{F}_Z)$

End

Output: HR- Hyperspectral image, Z

---

6. Results and discussion

In this paper, intially we implemented CNN-based fusion by extracting the spectral data from LR–HSI using 1D convolution network and spectral data from HR–MSI using 2D convolution network. These extracted spatial and spectral features are then fused together to obtain HR–HSI. To extract more detailed features from HS and MS, it requires deep CNN architecture. As CNN architecture become deeper, it introduced vanishing gradient problem. To overcome this, we implemented an unsupervised ResNet Fusion network by using skip connections. The proposed ResNet fusion inherits all the advantages of standard CNN. In addition to this, ResNet allows the designing of a deeper network without any performance degradation during the feature extraction process. Therefore, the proposed ResNet Fusion architecture extracts more discriminative features from both HSI and MSI and finally reconstruct a high-resolution HSI by fusing these extracted high-quality features from the ResNet.

The performance of CNN and ResNet fusion method is evaluated on four benchmark data sets using standard quality measures namely SAM, ERGAS, PSNR, and UIQI [30]. Further, we also compared the performance of CNN and ResNet fusion against the baseline fusion methods namely, CNMF [9], FCN-CNMF, and S2FEF-CNN [22]. Out of these, CNN shows better performance compared to CNMF and FCN-CNMF. The ResNet-based fusion shows outstanding performance compared to all other methods including CNN. The results obtained by CNN and ResNet fusion method against the baseline methods on four benchmark datasets are shown in **Table 2**. The low SAM indicates the good spectral data in the fused image and low ERGAS shows the statistical quality of the reconstructed image. The high PSNR and

Dataset	Methods	CNMF	FC-CNMF	CNN	S2FEF-CNN	ResNet
Pavia university	SAM	0.0633	0.0652	0.0451	0.0441	**0.0409**
	EARGAS	0.5423	0.4502	0.4311	0.4901	**0.4029**
	PSNR	64.4502	64.8923	65.1299	64.4915	**66.1127**
	UIQI	0.8779	0.9316	0.9262	0.9665	**0.9872**
Indian pines	SAM	0.5113	0.3976	0.4525	0.4118	**0.3896**
	EARGAS	0.8733	0.6991	0.6434	0.7192	**0.6170**
	PSNR	62.6779	63.1076	63.1311	64.8165	**65.2971**
	UIQI	0.7988	0.8432	0.8118	0.8776	**0.8991**
Washington DC mall	SAM	0.5609	0.5998	0.5956	0.5519	**0.5171**
	EARGAS	0.5741	0.5034	0.4993	0.4886	**0.4850**
	PSNR	64.09	64.12	64.19	65.11	**65.1358**
	UIQI	0.9199	0.9409	0.9213	0.9365	**0.9656**
Botswana	SAM	0.2541	0.2179	0.2233	0.2108	**0.1908**
	EARGAS	0.5194	0.4989	0.5034	0.4992	**0.4698**
	PSNR	63.1123	63.4321	63.9019	64.0116	**64.8798**
	UIQI	0.9703	0.9772	09715	0.9827	**0.9960**

Table 2.
The performance evaluation of different fused algorithms on four hyperspectral datasets.

UIQI show good spatial quality and high fidelity reconstructed image with less spectral distortion. From **Table 2**, it is further clear that good spectral preservation is obtained in Botswana dataset on analyzing the SAM value, which is reduced by more than 0.02 dB. Simultaneously, significant spatial preservation is achieved in the Indian Pine database revealed by the PSNR value increased by 1.5 dB.

The above work is extended by introducing different stacked convolution layers in the residual block of the ResNet. The experimental results obtained after stacked convolution layer in the ResNet are shown in **Table 3**. From the SAM value in **Table 3**, it is clear that the spectral information of the image is reducing as and when the number of stacked layers in the residual block increases. The UIQI value from the **Table 3** also reveals that quality of the reconstructed image is also diminishing as the number stacked layer increases in the ResNet. The PSNR and EARGS show a stable performance, which ensure the spatial consistency of our proposed method. So, we concluded that ResNet Fusion network with two-stacked convolution layer acquires more discriminative features from the source images and guarantee the quality of the reconstructed image on analyzing the results obtained in **Table 3**.

Figure 1 shown below is the visual representation of the output provided by our proposed ResNet fusion method on four benchmark datasets against all other baseline methods. From the figure, it is evident that ResNet Fusion with two-stacked convolution layers produces better performance in most of the areas in the image (highlighted) of the four datasets (**Figure 5**).

We further extend the Resnet fusion architecture to reduce the number of parameters to make our proposed method more efficient and effective to handle high

Dataset	Methods	Number of stacked convolution layers		
		2 layers	3 layers	4 layers
Pavia university	SAM	0.0409	0.065	0.069
	EARGAS	0.4029	0.4029	0.4029
	PSNR	66.1127	66.1127	66.1127
	UIQI	0.9872	0.9713	0.9622
Indian pines	SAM	0.3896	0.4186	0.4553
	EARGAS	0.6170	0.6170	0.6170
	PSNR	65.2971	65.2971	65.2971
	UIQI	0.8991	0.8904	0.8801
Washington DC mall	SAM	0.5171	0.5529	0.5721
	EARGAS	0.4850	0.4850	0.4850
	PSNR	65.1358	65.1358	65.1358
	UIQI	0.9656	0.9432	0.9209
Botswana	SAM	0.1908	0.1978	0.2085
	EARGAS	0.4698	0.4698	0.4698
	PSNR	64.8798	64.8798	64.8798
	UIQI	0.9960	0.9822	0.9589

Table 3.
The performance of ResNet fusion by varying the stacked layers.

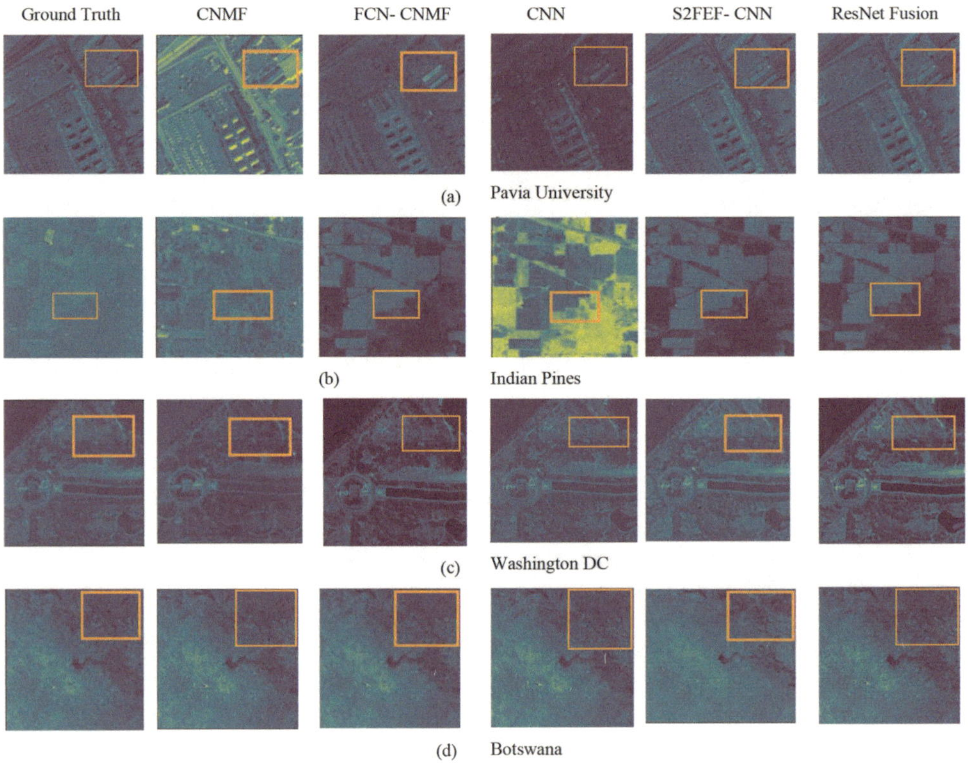

Figure 5.
The ground truth and fused image of different methods using four benchmark datasets.

Architecture	Number of parameters
CNN	31,586,081million
ResNet with Short Skip	8,045,825 million
ResNet with Long Skip	390,529 million
ResNet with Dense Skip	19,393 million

Table 4.
The performance of different skip connection.

dimensional data. For that, we used short skip, long skip, and dense skip connection to the ResNet architecture with two-stacked convolution layers. **Table 4** gives the total number of network parameters required for this ResNet architecture in each skip connection. From **Table 4**, it is clear that ResNet architecture with dense skip connection provides very less network parameters compared to ResNet with short and long skip connections.

A. Time complexity

Comparing the performance and running time of all the proposed algorithms on four benchmark datasets are shown in **Figure 6**. From this figure, it is evident that ResNet fusion with dense skip connection took very less running time and showed good performance in reconstructing high-fidelity hyperspectral image.

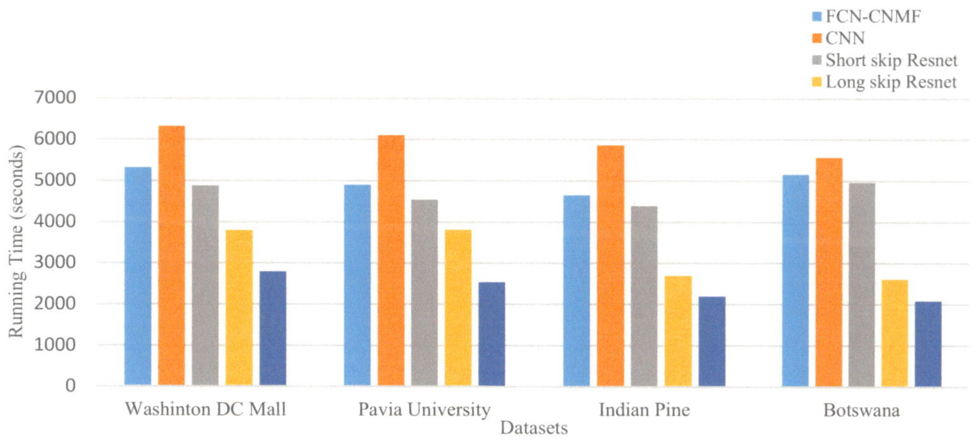

Figure 6.
The running time of traditional and deep learning HS-MS image fusion.

On comparing the performance and running time of ResNet with long skip and short skip connection, long skip connection ResNet fusion architecture shows good performance and running time than short skip connection. On evaluating the performance and running time of all ResNet fusion architectures, ResNet with dense skip connection outperformed compared to the other two ResNet fusion architectures. While comparing the performance and running time, the FCN-CNMF method showed better performance and time than CNN-based fusion. Finally, we concluded that, ResNet with dense skip connection with less network parameter shown highlighting performance for reconstructing good spatial and spectral quality HR-HSI compared to all other proposed methods. However, all our proposed methods show good in performance but the cost incurred in terms of time is high.

B. Resnet HS-MS fusion model

The experimental analysis of our ResNet fusion architecture with various parameters is done to build a general model for our proposed HS-MS ResNet fusion algorithm. For this purpose, we trained the network by using cropped HSI and MSI image pairs from each dataset. That means each dataset is cropped into several patches and then divided into training and testing data. In the case of Indian Pine dataset with size $610 \times 340 \times 103$ are cropped into several patches of size $M \times N \times L$. The patch size was $M\times N \times L = 15 \times 15 \times 103$ for Indian Pine dataset showing high performance to our network model. Similarly, we create training and testing samples for all three datasets. The patch size for Washington Dc Mall dataset was $M \times N \times L = 19 \times 19 \times 191$, for Botswana dataset, were $M \times N \times L = 17 \times 17 \times 145$ and for Pavia University dataset were $M \times N \times L = 19 \times 19 \times 192$ gives a network model with good running time and network parameters.

We measure the quality matrix value of our ResNet fusion by varying the number of stacked layers and found that residual blocks, each having two-stacked convolution layers is performing better than the others. The most significant part of ResNet is skipped connection, which helps for the flow of information through the network more efficiently and effectively. So, we also

Name	Layer	Kernel size	Input size	Input content	Stride	Padding	Activation	Output size	Output content
Input Layer	Conv 1	ID-CNN 1*3	1	1D Image(Spectral)	1	same	ReLU	64	1DConv1
		2D-CNN 3*3	2	2D Image(Spatial)	1	same	ReLU	64	2DConv1
Residual Block1	Conv 2	ID-CNN 1*3	64	1DConv1	1	same	ReLU	64	1DConv2
		2D-CNN 3*3	64	2DConv1	1	same	ReLU	64	2DConv2
	Conv 3	ID-CNN 1*3	64	1DConv2	1	same	ReLU	64	1DConv3
		2D-CNN 3*3	64	2DConv2	1	same	ReLU	64	2DConv3
Skip Connection	Add 1	—	—	1DConv1 + 1DConv3	—	—	—	—	1DResB1
				2DConv1+ 2DConv3					2DResB1
Residual Block2	Conv 4	ID-CNN 1*3	64	1DResB1	1	same	ReLU	64	1DConv4
		2D-CNN 3*3	64	2DResB1	1	same	ReLU	64	2DConv4
	Conv 5	ID-CNN 1*3	64	1DConv4	1	same	ReLU	64	1DConv5
		2D-CNN 3*3	64	2DConv4	1	same	ReLU	64	2DConv5
Skip Connection	Add 2	—	—	1DConv1 + 1DResB1 + 1DConv5	—	—	—	—	1DResB2
				2DConvt1 + 2DResB1 + 2DConv5					2DResB2
Residual Block3	Conv 6	ID-CNN 1*3	64	1DResB2	1	same	ReLU	64	1DConv6
		2D-CNN 3*3	64	2DResB2	1	same	ReLU	64	2DConv6
	Conv 7	ID-CNN 1*3	64	1DConv6	1	same	ReLU	64	1DConv7
		2D-CNN 3*3	64	2DConv6	1	same	ReLU	64	2DConv7
Skip Connection	Add 3	—	—	1DConv1 + 1DResB1 + 1DResB2 + 1DConv7	—	—	—	—	1DResB3
				2DConv1 + 2DResB1 + 2DResB2 + 2DConv7					2DResB3
Max pooling	Conv 8	ID-CNN 1*3	64	1DResB3	1	same	ReLU	32	1DConv8
		2D-CNN 3*3	64	2DResB3	1	same	ReLU	32	2DConv8

Name	Layer	Kernel size	Input size	Input content	Stride	Padding	Activation	Output size	Output content
Flatten layer	Conv 9	ID-CNN 1*1	32	1DConv8	1	same	ReLU	1	Spectral data
		2D-CNN 1*1	32	2DConv8	1	same	ReLU	1	Spatial data
Upsampling layer	Conv 10	2D-CNN 3*3	1	Spectral/Spatial data	1	same	ReLU	32	Spectral*Spatial
Output layer	Conv 11	2D-CNN 3*3	32	Spectral * Spatial	1	same	ReLU	64	Fused Image

Table 5.
ResNet-dense skip Architecture of HS-MS image fusion.

experimented with three skip connections: short skip, long skip, and dense skip connection. From this experiment, we found that ResNet with a dense skip connection reduces the number of network parameters to a large extent.

Finally, we built a generative ResNet model for the fusion of HS–MS image as shown in **Table 5**. The ResNet fusion model uses ID and 2D convolution networks. These two convolution networks consist of three residual blocks, each residual block contains two convolution layers with 64 filters, 3x3 kernel size, stride = 1, max-pooling, and padding = same. To make the information flow accurately throughout the network, we use dense skip connection. At last, it uses a 2D convolution to decode the reconstructed image into the original format.

7. Conclusion

In this work, we implemented HS–MS fusion on deep learning method because of its strong ability to extract features from the image. At first, we implemented the HS–MS fusion process in conventional CNN method. But in CNN, each layer takes the output from the previous layer, which tends to lose information as the network goes into deeper architecture. So we further implemented the fusion process in ResNet by adding the skip connection between the convolution layers. This skip connection helps to extract more detailed features from the images without any degradation problems. Our constructed ResNet fusion architecture includes three-residual blocks, and each block is a combination of stacked convolution layer and skip connections. Moreover, we modify the ResNet fusion architecture with different stacked layers and found that ResNet with two-stacked layer gives more accurate results. Finally, we extend ResNet architecture to reduce the number of parameters by using different skip connections like short ship, long skip, and dense skip connections. From the experimental analysis, it is found that the ResNet- dense skip improve the performance in image reconstruction with very less network parameters and running time compared to other fusion methods. This deep residual network helps to extract nonlinearity features with the help of the ReLU activation layer. The experiment and performance analysis of our algorithm is done effectively and quantitatively on four benchmark datasets. The fusion results indicate that ResNet with dense skip fusion method shows outstanding performance over traditional and DL methods by keeping the spatial and spectral data to a large extent in the reconstructed image.

Author details

K. Priya* and K.K. Rajkumar
Department of Information Technology, Kannur University, Kerala, India

*Address all correspondence to: kodothpriya@gmail.com

References

[1] Michael NH, Kudenov W. Review of snapshot spectral imaging technologies. Optical Engineering. 2013;**52**(10): 090901

[2] Feng F, Zhao B, Tang L, Wang W, Jia S. Robust low-rank abundance matrix estimation for hyperspectral unmixing. IET International Radar Conference (IRC 2018). 2019;**2019**(21): 6406-6409

[3] Dhore AD, Veena CS. Evaluation of various pansharpening methods using image quality metrics. 2nd International Conference on Electronics and Communication Systems (ICECS). IEEE. 18 June 2015:2015. DOI: 10.1109/ecs.2015.7125039

[4] Wang Z, Chen B, Ruiying L, Zhang H, Liu H, Varshney PK. FusionNet: "An unsupervised convolutional variational network for hyperspectral and multispectral image fusion". IEEE Transactions on Image Processing. 2020;**29**:7565-7577

[5] He K, Zhang X, Ren H, Sun J. Deep residual learning for image recognition. Proceedings of the IEEE conference on computer vision and pattern Recognition. IEEE. 12 December 2016: 770-778

[6] Loncan L, de Almeida LB, Bioucas-Dias JM, Briottet X, et al. Hyperspectral pansharpening: A review. In: IEEE Geoscience and Remote Sensing Magazine. IEEE; September 2015;**3**(3): 27-46

[7] Vivone G et al. A critical comparison among pansharpening algorithms. IEEE Transactions on Geoscience and Remote Sensing. 2015; **53**(5):2565-2586

[8] Wei Q, Bioucas-Dias J, Dobigeon N, Tourneret JY. Hyperspectral and multispectral image fusion based on a sparse representation. IEEE Transactions on Geoscience and Remote Sensing. 2015;**53**:3658-3668

[9] Patero and U. Tapper. Positive matrix factorization: A non-negative factor model with optimal utilization of error estimates of data values. Environmetrics. 1994;**5**:111-126

[10] Lee DD, Seung HS. Algorithms for non-negative matrix factorization. In: Advances in Neural Information Processing Systems. Denver. Cambridge, MA, United States: MIT press; 2001. pp. 556-562

[11] Tong L, Zhou J, Qian B, Yu J, Xiao C. Adaptive graph regularized multilayer nonnegative matrix factorization for hyper-spectral unmixing. IEEE Journal of Selected Topics in Applied Earth Observations and Remote Sensing. 2020;**13**:434-447

[12] Cao J et al. An endmember initialization scheme for nonnegative matrix factorization and its application in hyper-spectral unmixing. ISPRS International Journal of Geo-Information. 2018;**7**:195. DOI: 10.3390/ijgi7050195

[13] José M, Nascimento P, Bioucas Dias JM. Vertex component analysis: A fast algorithm to unmix hyperspectral data. IEEE Transactions on Geoscience and Remote Sensing. 2005;**43**(4)

[14] Yokoya N, Yairi T. Iwasaki, "A Coupled nonnegative matrix factorization unmixing for hyperspectral and multispectral data fusion". IEEE Transactions on Geoscience and Remote Sensing. 2012;**50**:528-537

[15] Simoes M, Bioucas-Dias J, Almeida L, Chanussot J. A convex formulation for hyperspectral image super resolution via subspace-based regularization. IEEE Transactions on Geoscience and Remote Sensing. 2015; **53**:3373-3388

[16] Lin C-H, Ma F, Chi C-Y, Hsieh C-H. A convex optimization-based coupled nonnegative matrix factorization algorithm for hyperspectral and multispectral data fusion. IEEE Transactions on Geoscience and Remote Sensing. 2018;**56**(3):1652-1667. DOI: 10.1109/tgrs.2017.2746078

[17] Yang F, Ma F, Ping Z, Guixian X. Total variation and signature-based regularizations on coupled nonnegative matrix factorization for data fusion. Digital Object Identifier. 2019;**7**: 2695-2706. DOI: 10.1109/ACCESS.2018. 2857943. IEEE Access

[18] Yang F, Ping Z, Ma F, Wang Y. Fusion of hyperspectral and multispectral images with sparse and proximal regularization. IEEE Access Digital Object Identifier. 2019;**2019**: 2961240. DOI: 10.1109/ACCESS

[19] Palsson F, Sveinsson JR, Ulfarsson MO. Multispectral and hyperspectral image fusion using a 3-D convolutional neural network. IEEE Geoscience and Remote Sensing Letters. 2017;**14**:639-643

[20] Masi G, Cozzolino D, Verdoliva L, Scarpa G. Pansharpening by convolutional neural networks. Remote Sensing. 2017;**8**(7):594

[21] Shao Z, Cai J. Remote sensing image fusion with deep convolutional neural network. IEEE Journal of Selected Topics in Applied Earth Observations and Remote Sensing. May 2018;**11**(5): 1656-1669

[22] Yang J, Zhao Y-Q, Chan J. Hyperspectral and multispectral image fusion via deep two-branches convolutional neural network. Remote Sensing. 2019;**10**(5):800

[23] Chen L, Wei Z, Xu Y. A lightweight spectral–spatial feature extraction and fusion network for hyperspectral image classification. Remote Sensing. 2020;**12**: 1395. DOI: 10.3390/rs12091395. 28 April

[24] Song W, Li S, Fang L, Lu T. Hyperspectral image classification with deep feature fusion network. IEEE Transactions on Geoscience and Remote Sensing. 2018;**56**(7):3173-3184

[25] Available from: http://lesun.weebly.com/hyperspectral-data-set.html

[26] Ian Goodfellow, Yoshua Bengio, Aaron Courville. Deep Learning. Available from: https://www.deeplearningbook.org/

[27] Ma F, Yang F, Ping Z, Wang W. Joint spatial-spectral smoothing in a minimum-volume simplex for hyperspectral image super-resolution. Applied Sciences. 2019;**10**(1)

[28] Available from: https://www.usgs.gov/landsat-missions/landsat-7

[29] Hong D, Yokoya N, Chanussot J, Zhu X. An augmented linear mixing model to address spectral varialbilty for hyperspectral unmixing, geography, computer science. In: IEEE Transactions on Image Processing. 2018

[30] Wang ACBZ. A universal image quality index. IEEE Signal Proessing Letters. 2002;**9**:81-84

Section 4

Future Trend of Hyperspectral Imaging Beyond Optical Domain

Chapter 5

Magnetic Scattering with Polarised Soft X-rays

Paul Steadman and Raymond Fan

Abstract

Soft X-ray scattering is a powerful technique for measuring magnetic materials. By highlighting some examples using diffraction, small angle scattering and reflectivity the element sensitivity and strong dependence of the polarisation on both the size and direction of the magnetic moments in both single crystals and thin films will be demonstrated.

Keywords: soft X-ray, magnetism, thin films, scattering, diffraction, reflectivity

1. Introduction

The interaction of light with magnetism was first discovered by Michael Faraday in 1845 when he observed that magnetised heavy glass would rotate the plane of polarised light as it was transmitted [1]. A few decades later John Kerr discovered the same magneto-optical effect but in a reflection geometry [2]. This proved the link between optics and magnetism, theoretically explained by James Clark Maxwell [3]. Whilst these first experiments were done using optical wavelengths [4–8] the first results using X-rays were not measured until 1972 (de Bergevin and Brunel) [9]. In this experiment, which was built on a previous idea (Platzmann and Tzoar 1970) [9] a laboratory X-ray source was used to measure the antiferromagnetic order in NiO. Several days were needed to collect the weak signal from the $(\frac{1}{4}\frac{1}{4}\frac{1}{4})$ peak due to the antiferromagnetic ordering between the main structural (charge) Bragg peaks.

This experiment was one of the first to prove that X-rays could be used to measure magnetism and that magnetic diffraction did not have to only rely on neutron diffraction. Indeed de Bergevin and Brunel neatly demonstrated that the interaction of both the electric and magnetic parts of X-ray. Unfortunately the interaction with the spin compared to the charge, is scaled by a relativistic factor of E_1/mc_2, where E_1 is the energy of the incident photon and m is the rest mass of the electron. This means that the magnetic scattering is reduced by a factor of 100 for each electron. Since not all electrons contribute to the magnetic signal the total magnetic signal is very weak compared to the scattering from structural(charge) Bragg peaks.

With investment in synchrotron radiation sources in the early 1980s, such as the SRS, Daresbury, UK or NSLS, New York, USA, the ability to separate the weak

 IntechOpen

magnetic scattering from the noise was increased by several orders of magnitude. With the high intensity of the synchrotron radiation and the well-defined polarisation meant that the effects discovered by de Bergevin and Brunel, which were weak and heavily polarisation dependent could be exploited. Magnetic scattering was now becoming a viable contender for measuring magnetism along with neutron scattering. The two techniques are actually very complimentary. The more bulk sensitive neutron scattering technique compares with a relative surface sensitive X-ray technique. An advantage of X-ray is the ability to be able to separate the spin and orbital parts of the electron angular momentum. This advantage is made possible through the different polarisation dependences of the scattering which had clearly been enhanced by synchrotron radiation.

A big breakthrough came at the end of the 1980s when Hannon et al. [10] discovered that magnetic scattering was enhanced at certain atomic resonances, in particular those from the dipolar transitions. Similarly to non-resonant scattering the spin and orbital parts of the electron could be separated. However, now the technique is element sensitive. The ability to access many energies on beamlines at synchrotrons enabled difference resonances or absorption edges to be accessed. In addition the dipole resonances enhancement are very strong at soft X-ray energies, which cover the $L_{2.3}$ resonances of the transition metals, the $M_{2,3}$ resonances of the rare earths and the N edges of the actinides.

In this chapter we will first discuss some theoretical preliminaries for resonant scattering, then soft X-ray diffraction followed by, small angle scattering, soft X-ray reflectivity and element specific hysteresis curves.

2. Theoretical preliminaries

X-ray magnetic scattering can be measured on or off an atomic resonance. The non-resonant scattering is stronger as the energy of the incident photon increases due to the relativistic factor mentioned previously. It is possible to measure X-ray magnetic scattering for energies above a keV (wavelengths of Angstroms) [11]. However, this is very weak at soft X-ray energies which are defined as energy between 100 and 2000eV (6.2 to 124 Angstroms). At both soft and hard X-ray energies magnetic scattering is enhanced by going to a resonance where a core electron with a well-defined spin (spin up or spin down) is transferred to the unoccupied states in the outer electron levels (same as the Fermi energy in metals). The well-defined spin then becomes a very sensitive probe of its environment which is short lived as it decays back to its core level emitting a photon of equal energy to the incident one (elastic scattering). In the dipole approximation the spin does not flip so spin is preserved throughout this process which make it very sensitive to the magnetic moment of the atom, since the outer electron levels is the magnetic environment. In addition the magnetic order breaks the symmetry of the lattice, since this is a vector quantity thus any experiment involving this resonant process i.e. X-ray absorption or X-ray scattering has strong polarisation dependence.

We will not discuss non resonant scattering but there are reviews in the literature [12–14] as well as some of the first work by De Bergevin, Brunel, Gibbs and Blume to name but a few [9, 12].

The amplitude of electric dipole transitions can be written as [10, 16].

$$f = (\mathbf{e_f} \cdot \mathbf{e_i})\ F^{(0)} - i(\mathbf{e_f} \times \mathbf{e_i}) \cdot \mathbf{M}F^{(1)} + (\mathbf{e_f} \cdot \mathbf{M})(\mathbf{e_i} \cdot \mathbf{M})F^{(2)} \quad (1)$$

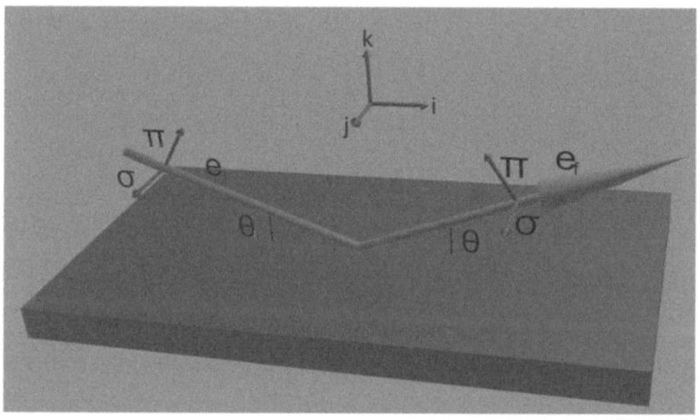

Figure 1.
The frame of reference used for the calculations of polarisation dependent scattering. The Greek symbols π and σ refer to polarisation that are parallel or perpendicular to the scattering plane (plane defined by incoming and outgoing beam) respectively. The suffixes i and f refer to the incident and scattered polarisation. The incident and outgoing angles are represented by θ_i and θ_f respectively. A right handed set with unit vectors i, j and k is shown.

Here $\mathbf{e}_i$ and $\mathbf{e}_f$ are directional vectors representing the incident and scattered polarisation respectively, **M** is the magnetic moment and the coefficients $F^{(0)}$, $F^{(1)}$and $F^{(2)}$ depend on the matrix elements involved in the resonant process. The discussion of these coefficients are out of the scope of this work and not necessary for this chapter or our conclusions but some comments will be necessary. The first term is the resonant charge scattering and has the same form as the non-resonant charge scattering here the dot product between the polarisation vectors is simply due to experimental geometry i.e. the position of the detector relative to the incident beam. As the detector angle increases the dot products of the polarisation vectors components in the scattering plane has a cosine dependence and therefore gets weaker. The second term is a first order in magnetic moment. It involves a triple product with the cross product between the two polarisation vectors dotted with the magnetic moment. This term is related to circular dichroism in absorption. The third term is the second order in magnetic moment and depends on the projection of each polarisation vector with the magnetic moment. It gives raise to the linear dichroism in absorption. When referring to the above we will refer to the reference frame shown in **Figure 1**. Here we define the polarisation vectors π_i and π_f in the scattering plane and σ_i and σ_f perpendicular to the scattering plane for incoming and outgoing polarisations respectively. With this in mind and defining θ_i and θ_f as the incoming and outgoing grazing angles we can define the following vectors.

$$\boldsymbol{\pi}_\mathbf{i} = \pi_i \sin(\theta_i)\mathbf{i} + \pi_i \cos(\theta_i)\mathbf{k} \tag{2}$$

$$\boldsymbol{\sigma}_\mathbf{i} = \sigma_i \mathbf{j} \tag{3}$$

$$\boldsymbol{\pi}_\mathbf{f} = -\pi_f \sin(\theta_f)\mathbf{i} + \pi_f \cos(\theta_f)\mathbf{k} \tag{4}$$

$$\boldsymbol{\sigma}_\mathbf{f} = \sigma_f \mathbf{j} \tag{5}$$

With this frame of reference we would like to construct the following matrix equation where each element represents a well-defined initial and final polarisation state.

$$f = \begin{pmatrix} \sigma_i \to \sigma_f & \pi_i \to \sigma_f \\ \sigma_i \to \pi_f & \pi_i \to \pi_f \end{pmatrix} F^{(0)} - i \begin{pmatrix} \sigma_i \to \sigma_f & \pi_i \to \sigma_f \\ \sigma_i \to \pi_f & \pi_i \to \pi_f \end{pmatrix} F^{(1)} + \begin{pmatrix} \sigma_i \to \sigma_f & \pi_i \to \sigma_f \\ \sigma_i \to \pi_f & \pi_i \to \pi_f \end{pmatrix} F^{(2)} \tag{6}$$

If we assume that θ_i and θ_f are equal to θ as in specular reflectivity and use Eqs. (2)–(5) one can rewrite the Eq. (1)

$$f = \begin{pmatrix} 1 & 0 \\ 0 & \cos 2\theta \end{pmatrix} F^{(0)} - i \begin{pmatrix} 0 & m_i \cos\theta - m_k \sin\theta \\ -m_i \cos\theta - m_k \sin\theta & m_j \sin 2\theta \end{pmatrix} F^{(1)} + \begin{pmatrix} m_k^2 & m_k(m_i \sin\theta + m_k \sin\theta) \\ -m_k(m_i \sin\theta - m_k \sin\theta) & -\cos^2\theta\left(m_i^2 \tan^2\theta + m_k^2\right) \end{pmatrix} F^{(2)} \tag{7}$$

Where

$$\mathbf{M} = m_i\hat{\mathbf{i}} + m_j\hat{\mathbf{j}} + m_k\hat{\mathbf{k}} \tag{8}$$

Although this is just another version of Eq. (1) in a particular frame of reference, it makes it easier to see that the off-diagonal components within the first order term only depend on the magnetic moment within the scattering plane i.e. m_i and m_k and the diagonal term only depends on the magnetic moment out of the scattering plane m_j. The second order term in magnetic moment is more complicated and allows magnetic scattering in the $\sigma_i \rightarrow \sigma_f$ channel of the matrix. The second order terms matrix tend to be small so we will ignore this for most of the chapter but some comments will be made on this when we discuss diffraction.

We will now apply these equations in three different situations. In the next section we will briefly examine the subject of diffraction, then small angle scattering and followed by a section dedicated to reflectivity measurements.

3. Diffraction

There are many exciting materials with large enough unit cells to enable the Bragg condition to be satisfied at soft X-ray wavelengths. In addition since magnetism lowers the symmetry of the crystal lattice, it is possible that extra diffraction peaks will occur in between the main Bragg peaks, due to the larger magnetic unit cell. This can help enormously with soft X-ray scattering since even if it is not possible to reach one of the main Bragg peaks it may be possible to reach a magnetic diffraction peak.

In kinematical theory we sum up the diffraction amplitudes as follows

$$A(\mathbf{Q}) = \sum_{n=o}^{N-1} f_n \exp\left(i\mathbf{Q}.\mathbf{r}_n\right) \tag{9}$$

where f_n is the form factor of a particular element, $\mathbf{q}$ is the reciprocal lattice vector and $\mathbf{r}_n$ is the real space position of the atoms in the lattice. In the case where the scattering ion is at resonance we need to replace the form factor f_n with the anomalous corrections as shown in Eq. (10).

$$f \rightarrow f + f' + if'' - if_m \tag{10}$$

where the f is the non-dispersive atomic form factor, f' and f'' are the real and imaginary parts of the dispersion corrections to the charge scattering and f_m is the contribution from the magnetic scattering.

3.1 Commensurate antiferromagnet

The system $Co_xMn_{1-x}WO_4$ [17] which has many different phases one of which is an interesting multiferroic phase at low temperatures, another is a commensurate antiferromagnetic structure. The magnetic moments on the Mn atoms in this phase align antiparallel along the **a** direction. The lowering of the crystal symmetry means that the unit cell is doubled compared to the chemical unit cell. This means that in between the chemical Bragg peaks there are peaks at "half order" positions that are purely magnetic. This is demonstrated with a simple schematic in **Figure 2**.

The antiferromagnetic phase in $Co_xMn_{1-x}WO_4$, known as the AF4 phase, exists in samples with $x = 0.15$ below 18 K. So by going to the Co or Mn $L_{2,3}$ resonance and then putting the sample and detector at the correct point in reciprocal space we should be able to measure the Brag diffraction due to the AF4 antiferromagnetic order. This is shown in 2 where we see the resonance enhancement appears very clearly at the Co L_3 edge at the $(\frac{1}{2}\ 0\ 0)$ position. One of the advantages of X-ray resonant scattering is the ability to tune and distinguish between different elements. This is nicely demonstrated in this sample where it was possible to tune to the Co L_3 and Mn L_2 edges and follow their evolution with temperature. Both peaks behave similarly and decay as expected from previous work. Whilst this demonstrates the power of soft X-ray scattering, and X-ray scattering in general, particularly with element specificity it is worth noting the disadvantages. Firstly having the half order peak was necessary so that there was a peak existing in reciprocal space that could be measured i.e. where the Bragg condition could be satisfied. Also since the Mn resonances occur at lower energy right at the limit of where the Bragg condition could be satisfied. From this particular sample, it was only possible to measure the $(\frac{1}{2}\ 0\ 0)$ peak at the Mn L_2 edge since the Mn L_3 edge, occurring at 638.7 eV, was too low in energy.

Another tool one can use in soft X-ray scattering is polarisation analysis. By looking at the form of Eq. (1). In particular the first order in magnetic moment, the triple product $(\mathbf{e_f} \times \mathbf{e_i}) \cdot \mathbf{M}$ involves the incoming polarisation $\mathbf{e_i}$, the outgoing polarisation $\mathbf{e_f}$ along with the magnetic moment **M**. This vector nature of this process means that if one is able to define the incoming polarisation, measure the outgoing polarisation along with its intensity then it should be possible to measure the direction of the magnetic moment. Since there are an infinite number of solutions if the measurement is only done using one incident polarisation, a technique known as full polarisation analysis is used. Here several incoming polarisations are used and then the outgoing polarisation is measured for each one. The outgoing polarisation is measured by taking

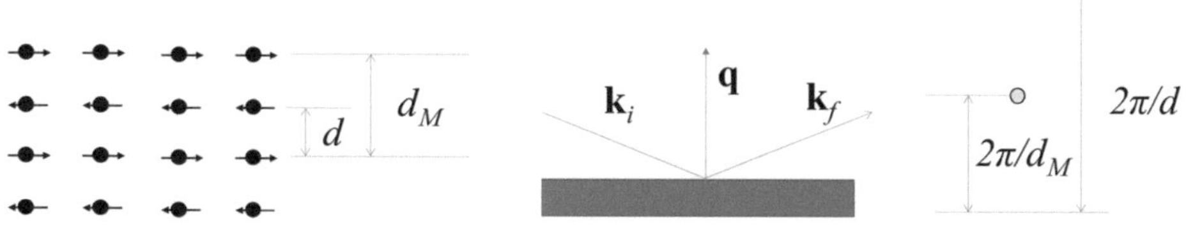

Figure 2.
A schematic demonstrating how doubling the size of the unit cell, shown on left, results in half order peaks in reciprocal space, shown on right.

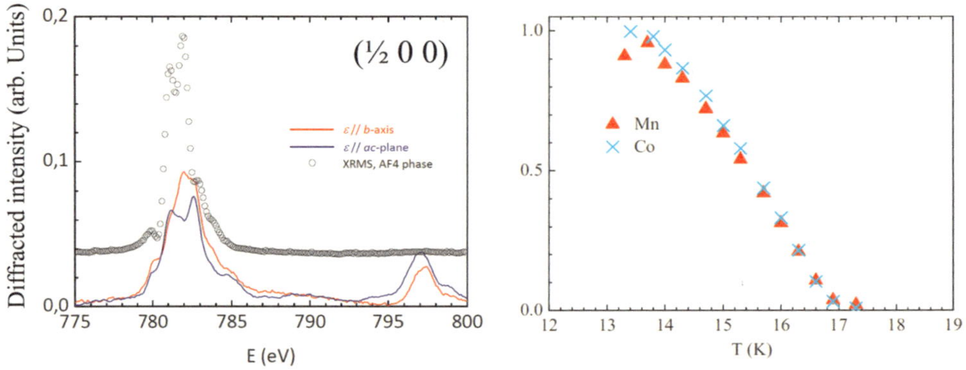

Figure 3.
The ($\frac{1}{2}$ 0 0) peak of $Co_{0.15}Mn_{0.85}WO_4$. On the left is shown the intensity (symbols) of the peak as one changes the energy through the Co $L_{2,3}$ resonances. Also shown are the fluorescence signals (line) as a function of energy in two different crystalline directions. On the right is the intensity at the Co L_3 and Mn L_2 edge as a function of temperature.

the scattered beam and diffracting it at 90 ° using a special multilayer analyser. With a specially designed detector mount the scattered beam polarisation is measured by rotating the detector and analyser around the scattered beam. The results of doing this at the Co L_3 edge are shown in **Figure 3**. By fitting the $(\mathbf{e_f} \times \mathbf{e_i}) \cdot \mathbf{M}$ for the outgoing polarisations for several incoming polarisations it was possible to determine the direction of the magnetic moments. The resonant nature of this scattering also meant that it was possible to ascertain that the Mn and Co moments are non collinear, a measurement that would not have been possible with other techniques. The non collinearity, a result of competition between the Co^{2+} and Mn^{2+} single ion anisotropies furthers the understanding of the complex magnetic phase diagram of this material (**Figure 4**) [18].

3.2 Incommensurate structures

In addition to commensurate magnetic lattices there are examples of magnetic lattices that are incommensurate with the chemical structure. Such structures still provide diffraction peaks as can easily be shown in the following example. If we take Eq. (9) for a one dimensional lattice and add in an incommensurate modulation in the magnetic moments similar to an example shown in [19] (see section 4.4.5) but adapted to magnetism.

$$A(\mathbf{Q}) = \sum_{n=o}^{N-1} \left(f + f' + if''\right) \exp\left(iQ.r_n\right) - if^m \exp\left(iQ.r_n\right) \exp\left(iq_m.r_n u \cos\left(q.r_n\right)\right) \tag{11}$$

In this equation we have assumed a complex atomic form factor $f + f' + if'' - if_m$ where f is the non-dispersive form factor with the real and imaginary terms of the dispersive form factors f' and f'' respectively. There is also an additional part f_m due to the magnetic moment which includes all the magnetic terms in 1. The second order term, however, will be assumed to be negligible. The term u is the amplitude of a wave

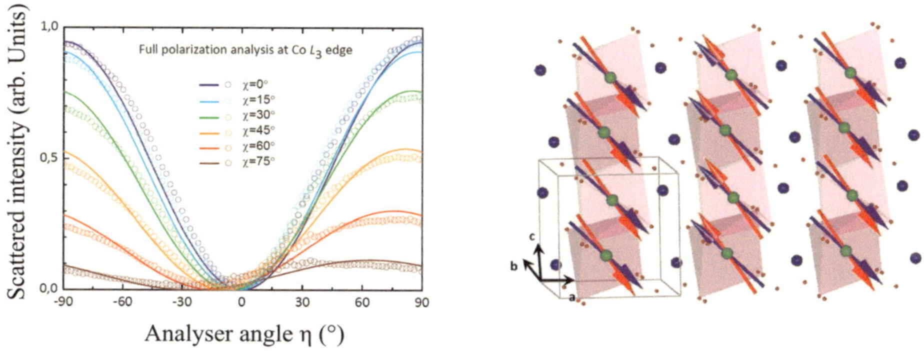

Figure 4.
On the left is shown the full polarisation analysis measurement at the Co edge. For each angle of incident linear polarisation given by χ the analysis of the polarisation was performed by rotating the detector analyser around the scattered beam (rotation of detector analyser angle η). A value of zero is defined as perpendicular to the scattering plane. By doing this at both the Co L_3 and Mn L_2 edges it was possible to measure both moments. The results are depicted in the picture on the right where it was established that the Co and Mn moments were not parallel.

which forms the incommensurate structure with a periodicity given by $q_m = 2\pi/\lambda_m$. We can expand the exponential containing the modulation such that:

$$A(\mathbf{Q}) = \sum_{n=o}^{N-1}\left(f + f' + if''\right)\exp\left(iQ.r_n\right) - if^m \exp\left(iQ.r_n\right)\left(1 + iQu\cos\left(q_m.r_n\right) + \ldots\right) \tag{12}$$

To first order gives.

$$A(\mathbf{Q}) = \sum_{n=o}^{N-1}\left(f + f' + if'' - if_m\right)\exp\left(iQ.r_n\right) + f^m\left(\frac{Qu}{2}\right)\left[\exp\left(i\left(Q + q_m\right).r_n\right)\right.$$
$$\left. + \exp\left(i\left(Q - q_m\right).r_n\right)\right] \tag{13}$$

By writing $r_n = an$, where a is the lattice parameter of the one dimensional lattice we can work out the summations. The modulus squared then gives us the intensity which, in the limit of large N, yields the following.

$$I(\mathbf{Q}) = N\left(\frac{2\pi}{a}\right)\left[\left(f + f'\right)^2 + \left(f'' - f_m\right)^2\right]\delta(Q - G) \tag{14}$$
$$+ N\left(\frac{Qu}{2}\right)^2\left(\frac{2\pi}{a}\right)f_m{}^2\left[\delta\left(Q + q_m - G\right) + \delta\left(Q - q_m - G\right)\right]$$

Here the δ are the Dirac delta functions and the G are reciprocal lattice vectors along the one dimensional lattice. This means that as well as the structural Bragg peak $Q = G$ the modulated magnetic structure gives magnetic peaks around the Bragg peak at $Q = G \pm q_m$. In this example there are only first order peaks but that is because we only took the expansion in Eq. (12) to first order. Note that magnetic scattering is also presence at the Bragg peak and not just on the peaks around it. This can be seen by the f_m in the factor of the Dirac delta function for the main Bragg peak.

There are many fascinating example of incommensurate magnetic structures. Hexaferrites, an interesting materials with multiferroic properties offer interesting properties to study with soft X-rays [20, 21]. The large unit cells of the M, Y and Z type hexaferrites enable the Bragg condition to be satisfied even at soft X-ray energies (particularly at the Fe and Co $L_{2,3}$ edges). Incommensurate spin structures result in easily reachable magnetic diffraction peaks which can be studied at different temperatures, magnetic and electric fields.

4. Small angle scattering

Another possibility to measure magnetic structures is to perform experiments in transmission enabling the measurement of small angle scattering. Due to the strong absorption of soft X-rays the samples have to be about a few hundred nanometres thick or thinner. The complexity of producing the samples is a contrast to the much simpler experimental set-up. Since the energies are quite low there is the opportunity to study large structures such as magnetic domains. A very good example of this is the study of the domains in FeRh with both circular and linear polarisation [22]. In this work the domains and their evolution over time across the interesting antiferromagnetic to ferromagnetic transition was examined. Another area that has made extensive use of small angle scattering involves magnetic skyrmions. Magnetic skyrmions can best be described as textures of magnetic swirls. They are caused by a balance of magnetic anisotropy, applied field, fluctuating temperature and the Dzyaloshinskii-Moriya interaction. The latter, caused by the electronic spins sensitivity to non-centro-symmetric symmetry via the spin-orbit interaction causes the magnetic spins to spiral in two dimensions (see **Figure 5**). The topological nature of the spin structure means that they are robust magnetic entities which could potentially be used in magnetic memory applications [23].

A typical phase diagram of magnetic states in a skyrmion hosting material is shown in **Figure 5**. In general in the absent of magnetic field there is a helical arrangement of spins. If a field is applied the spins start to rotate towards the applied field. At certain values of applied field and temperature the skyrmion phase occurs. The exact values

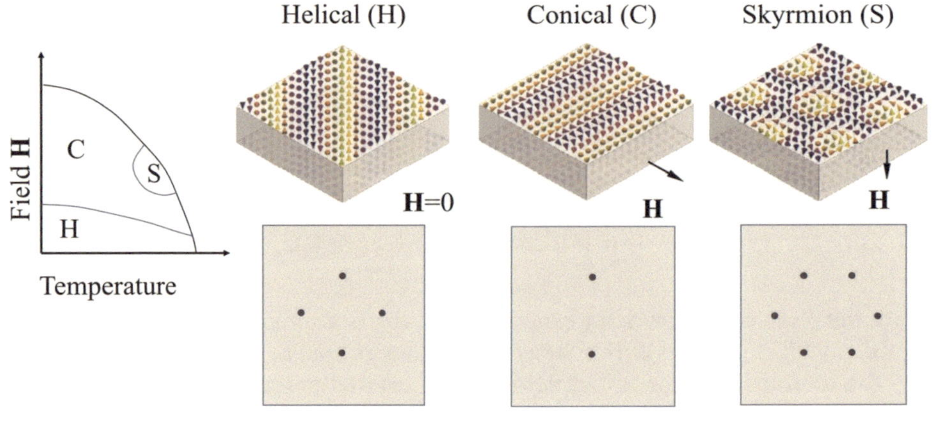

Figure 5.
The different phases that exist in skyrmion hosting materials. On the left is shown a schematic of a phase diagram along with pictures of the helical, conical and skyrmion phases. The corresponding diffraction patterns are shown below for each phase.

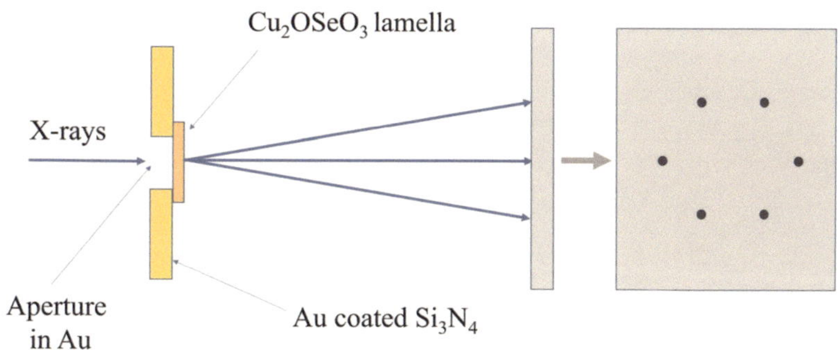

Figure 6.
The setup for small angle scattering. The X-rays pass through an aperture and are transmitted through an ultrathin sample. The small angle scattering is measured using a CCD camera. A schematic of the hexagonal diffraction pattern from a skyrmion lattice is shown on the right.

of temperature and magnetic field that this phases occur depends on the material and more specifically on the exchange interaction, Dzyaloshinskii-Moriya interaction, spin-orbit interaction and crystalline anistropy. Also shown are the typical in **Figure 5** are the diffraction patterns due to the scattering from skyrmions and the competing helical and conical phases.

The large magnetic periodicity of the skyrmion lattices, which can vary from tens to hundreds of nanometres makes them ideal for soft X-ray diffraction. Many experiments have been done on Cu_2OSeO3_3 which, unlike some of the other B20 materials, is insulating. Here the skyrmion lattice causes a six fold diffraction pattern around the (0 0 1) diffraction peak. This occurs in a similar way to the commensurate antiferromagnet mentioned previously but the magnetic lattice is now many times larger than the chemical unit cell of the Cu_2OSeO3_3 (see for example [24–26]).

Another way of measuring skyrmions is to grow very thin samples and measure the small angle scattering in transmission. The technique of small angle neutron scattering (SANS) has already been used extensively for measuring skyrmions (e.g [27, 28]). The hexagonal structure of the skyrmion lattice will produce a hexagonal diffraction pattern around the (0 0 0) incident beam direction. A schematic is shown in **Figure 6**.

An example of such measurements using small angle scattering is shown in **Figure 7** where skyrmions were measured on thin samples of Cu_2OSeO3_3 [29]. Here it was demonstrated that by field cooling (in a field of 44 mT) the skyrmion phase existed all the way down to 23.5K. Moreover there is no anomaly at the measured phase boundary (40K) as can be seen in the evolution of the reciprocal lattice vector and the intensity of the diffraction peaks. The intensity was fitted with a power law and gave a critical exponent of 0.73 which is expected for a three dimensional Heisenberg system agreeing with previous work [30, 31].

5. Reflectivity

To avoid ambiguity reflectivity in this chapter will refer to the case of specular reflectivity i.e. where the incoming angle is equal to the outgoing angle. A reflectivity scan is generally performed by increasing the detector angle at twice the rate of the sample angle although some commercial diffractometers allow the symmetric

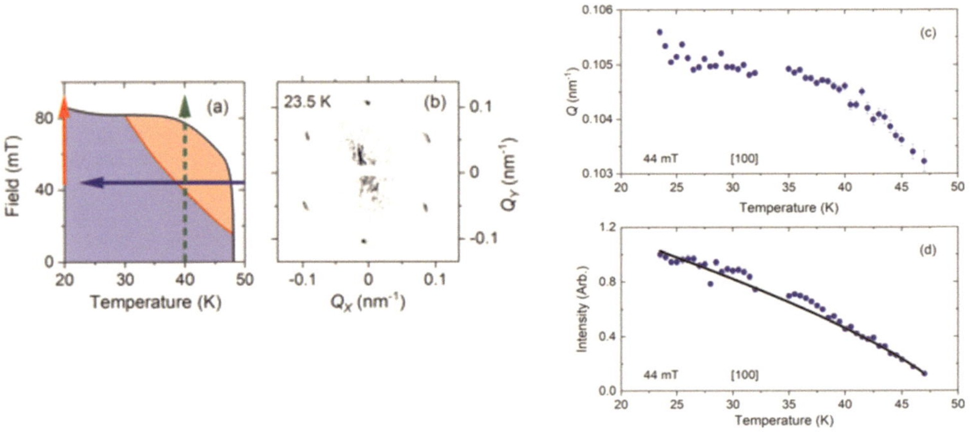

Figure 7.
The production of metastable skyrmions. The blue line in the phase diagram (a) results in a metastable skyrmion phase (b). (c) and (d) show the evolution of the reciprocal lattice vector Q and the intensity of the skyrmion peaks. The solid line in d is a fit to a power law behaviour yielding a critical exponent of 0.73.

increasing of the incoming and outgoing angle by increasing the detector and X-ray angles but keeping the sample constant.

Although it is a scattering technique like diffraction, reflectivity is different. Whilst diffraction refers to scattering from planes of atoms, reflectivity refers to scattering from a surface or interface or a combination of both. In many cases diffraction can be described by kinematical theory where amplitudes can be summed up. Reflectivity is often best described by optical theories using the Fresnel coefficients for reflection from each interface. In soft X-ray reflectivity this usually works well since the wavelengths are large enough to assume that the material is a continuum and not discrete planes atoms (as in diffraction). However, if a Bragg condition is satisfied during a reflectivity measurement (which would be quite common in a hard X-ray measurement) then the optical theory will no longer adequately describe the scattering and more complicated dynamical theories are needed [32].

An example of an optical theory that works well with soft X-ray scattering involves that of Zak et al [32–35]. It involves calculating the wave properties as it propagates through a multilayer system. Two matrices are formulated: one that calculates the electromagnetic waves due to the reflection and refraction at each interfaces and a second one calculates the phase of the wave. The details are included in the references. Although it is based on optical theory, for calculating the Kerr and Faraday rotations it works well for soft X-rays as long as there is not have any Bragg diffraction i.e. that we can model the films as continuous media. It is a classical equivalent of the theory represented by Eq. (1) to first order in magnetisation.

Soft X-ray reflectivity is a very powerful technique for studying thin films and multilayers and therefore very relevant for device applications. A good example is exchange bias. Exchange bias occurs when a ferromagnetic is grown next to an antiferromagnetic material. The coupling at the interface causes a unidirectional anistropy; a hysteresis loop of the ferromagnetic material is not centred at zero applied field but offset by a quantity known as the exchange field H_{ex}. Discovered in 1956 by Meiklejohn and Bean [36] the effect is still not properly understood despite being used in read heads in the latest hard drives. $IrMn_3$ is an antiferromagnet and the most commonly used in spin valves in hard drives. A layer of Fe grown on top of $IrMn_3$ is exchange biased. The nature of the reversal process and the exchange bias field H_{ex} depends strongly on the structure of the antiferromagnet and therefore on its method of preparation. $IrMn_3$ can be prepared by molecular beam epitaxy or the more industrially relevant sputtering. It is well known that when $IrMn_3$ is grown at room temperature the structure is a γ phase where the atoms are arranged randomly in a face centred cubic structure. When it is grown with the substrate at 400°C the material orders in the $L1_2$ structure. Here the structure can be described as having the Mn atoms ordered in the centre of the faces giving a simple cubic structure. In addition an Fe film grown on top of this will have a very different magnetic reversal behaviour. The hysteresis loops are shown in **Figure 8**.

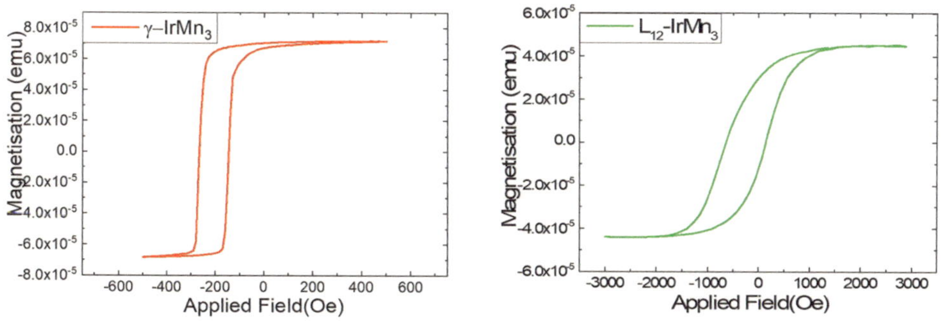

Figure 8.
The hysteresis curves from $IrMn_3$. The loop for the γ disordered phase is shown on the left and that corresponding $L1_2$ phase to the right.

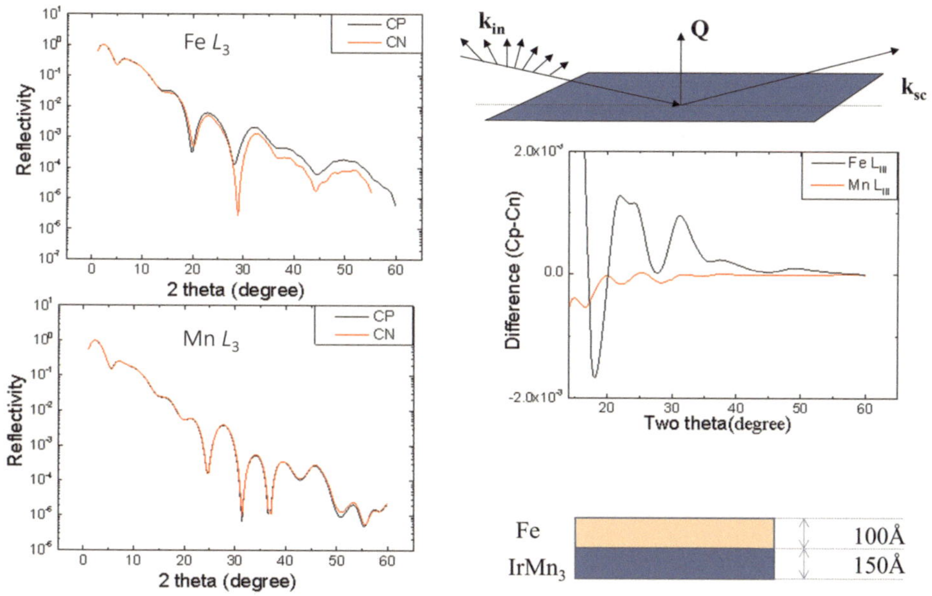

Figure 9.
Reflectivity measure for the disordered γ phase of $IrMn_3$. On the right are shown the reflectivities measured at the Fe L_3 edge (top) and Mn L_3 edge at the bottom. The reflectivities were measured with incident circularly polarised light at both helicities indicated by CN and CP in the plots. A schematic of the measurement geometry is shown at the top right. The plot on the right is the difference between the two helicities (dichroism). A schematic of the thin film with the thicknesses of the Fe and $IrMn_3$ layers is shown at the bottom right.

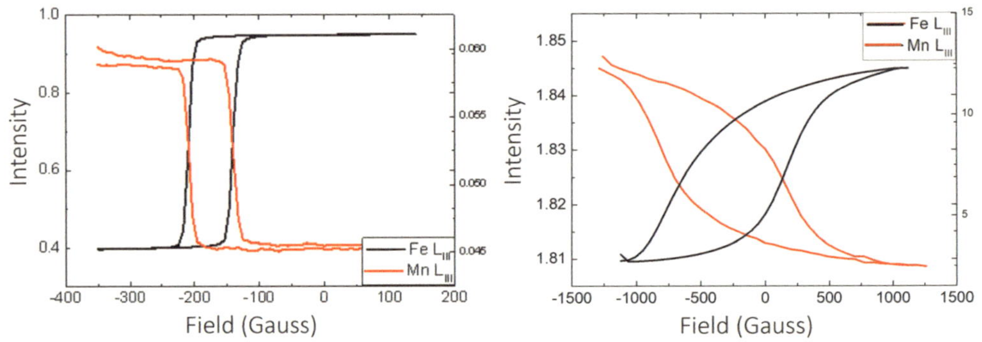

Figure 10.
Magnetic reflectivity measured during a hysteresis cycle. This was achieved by measuring at the Fe and Mn L_3 edges (Fe in black and Mn in red). On the left are shown the results from disordered γ phase of $IrMn_3$. The corresponding ones from the $L1_2$ ordered phase are shown on the right.

It can immediately be seen from **Figure 8** that both hysteresis loops are exchange biased. However, the γ phase has a sharp loop with a H_{ex} of 150 Oe, the $L1_2$ sample is much smoother and with a much higher H_{ex}. Understanding the mechanisms for this is vital for understanding exchange bias and improving spintronics devices.

By tuning to the $L_{2,3}$ resonances of Mn and Fe it is possible to separate out the magnetism from the ferromagnetic and the antiferromagnetic layer. By using circular polarisation the technique also benefits from the significant linear component of magnetisation in the scattering cross section. This is shown in **Figure 9**. On the left is shown two reflectivities; the top one is measured at the Fe L_3 resonance (sensitive to the ferromagnet) and the bottom one is measured at the Mn L_3 edge (sensitive to the antiferromagnet). In each reflectivity opposite helicities of circular polarisation were measured. These are designated CP and CN in the plots. At the Fe L_3 edge there is a clear difference between the two reflectivities measured at opposite helicities which is not so apparent in the reflectivities measured at the Mn L_3 edge. This is represented in the plot of the difference on the right. Here the difference, often called the dichroism, is measured. This is not to be confused with the magnetic circular dichroism of X-ray absorption although it is strongly related. It is noteworthy here that whilst there is a clear difference at the Fe edge, hardly surprising for a ferromagnetic material, there is also a small but significant difference at the Mn edge.

To examine the magnetic signal more we could fix the sample and detector angles at a convenient point in reciprocal space and measure the intensity as the sample goes through a hysteresis cycle. The result of this measurement is shown in **Figure 10**. Here it can clearly be seen that the signal follows the hysteresis much like that produced by a vibrating sample magnetometer. With X-rays we have the added advantage of being element specific which is nicely demonstrated here; by tuning to the Fe resonance we are measuring the ferromagnetic behaviour and at the Mn resonance we are measuring the antiferromagnet.

It is hardly surprising that we can measure the ferromagnet. The terms in the first order (in magnetic moment) part of Eq. (7) show that magnetic scattering measures the magnetic moment in several directions depending on the magnetic moment.

To show this we can write out Eq. (7) in the following way

$$f = \begin{pmatrix} \sigma_{CR11} & 0 \\ 0 & \pi_{CR22} \end{pmatrix} + i \begin{pmatrix} \sigma_{CI11} & 0 \\ 0 & \pi_{CI22} \end{pmatrix} - i \begin{pmatrix} 0 & \sigma_{M12} \\ \pi_{M21} & \pi_{M22} \end{pmatrix} \quad (15)$$

Here we have added an imaginary term for the charge scattering to allow for the phase change during the resonant process. The imaginary term is assumed to have the same polarisation dependence as the real term. The magnetic term is assumed only to have an imaginary part. We have also ignored the second order part of the equation which we assume to be negligible. For circular polarisation we need to construct the polarisation as two orthogonal components with a π phase difference i.e.

$$\begin{aligned} P_i^+ &= \sigma_i + i\pi_i \\ P_i^- &= \sigma_i - i\pi_i \end{aligned} \tag{16}$$

For both helicities respectively. Here the + and - refer to the different helicities of the circular polarisation. Including now the phase factors the structure factors for both helicities become

$$\begin{aligned} f^+ &= \begin{pmatrix} \sigma_{CR11} & 0 \\ 0 & i\pi_{CR22} \end{pmatrix} + i\begin{pmatrix} \sigma_{CI11} & 0 \\ 0 & i\pi_{CI22} \end{pmatrix} - i\begin{pmatrix} 0 & i\sigma_{M12} \\ \pi_{M21} & i\pi_{M22} \end{pmatrix} \\ f^- &= \begin{pmatrix} \sigma_{CR11} & 0 \\ 0 & -i\pi_{CR22} \end{pmatrix} + i\begin{pmatrix} \sigma_{CI11} & 0 \\ 0 & -i\pi_{CI22} \end{pmatrix} - i\begin{pmatrix} 0 & -i\sigma_{M12} \\ \pi_{M21} & -i\pi_{M22} \end{pmatrix} \end{aligned} \tag{17}$$

With this we can work out the scattered intensity. Here we work out a general expression with the applied magnetic field along any direction.

$$\begin{aligned} I^+ &= F_{Total}F_{Total}^* = F_C F_C^* + F_C F_M^* + F_M F_C^* + F_M F_M^* \\ &= \sigma_{CR11}^2 + \pi_{CR22}^2 + \sigma_{CI11}^2 + \pi_{CI22}^2 + \sigma_{M12}^2 + \pi_{M21}^2 + 2(\sigma_{CR11}\sigma_{M12} - \pi_{CR22}\pi_{M21}) \\ &\quad + \pi_{M22}^2 - 2\pi_{CI22}\pi_{M22} \\ I^- &= F_{Total}F_{Total}^* = F_C F_C^* + F_C F_M^* + F_M F_C^* + F_M F_M^* \\ &= \sigma_{CR11}^2 + \pi_{CR22}^2 + \sigma_{CI11}^2 + \pi_{CI22}^2 + \sigma_{M12}^2 + \pi_{M21}^2 - 2(\sigma_{CR11}\sigma_{M12} - \pi_{CR22}\pi_{M21}) \\ &\quad + \pi_{M22}^2 - 2\pi_{CI22}\pi_{M22} \end{aligned} \tag{18}$$

We note here that the π phase difference in front of the magnetic terms means that it is possible to have interference between the charge and magnetic scattering, interference terms which are linear in σ_{M12}, π_{M21} and π_{M22}. Note also that these terms change sign with helicity. However there are also quadratic terms in σ_{M12}, π_{M21} and π_{M22}. As the field is being applied in the scattering plane the π_{M22} terms, which only depend on moments out of the scattering plane are small. This will work reasonably for the disordered sample where the moments switch abruptly. If we were to ignore the pure charge scattering terms then the two terms above can be simplified to the following.

$$\begin{aligned} I^+ &= \sigma_{M12}^2 + \pi_{M21}^2 + 2(\sigma_{CR11}\sigma_{M12} - \pi_{CR22}\pi_{M21}) \\ I^- &= \sigma_{M12}^2 + \pi_{M21}^2 - 2(\sigma_{CR11}\sigma_{M12} - \pi_{CR22}\pi_{M21}) \end{aligned} \tag{19}$$

We can now see that the effect of changing the helicity during a scattering measurement of a hysteresis would resulted in the reverse the loop. However, the quadratic terms cannot always be ignored. Since the quadratic terms obviously do not reverse with helicity a simple way of removing this uncertainty is to measure the scattering during hysteresis with opposite helicities and subtract one from the other i.e. take the dichroism of the measured hysteresis. The important result from Eq. (19) is that there is a linear dependence on magnetic moment which reverses with helicity explaining why we see the hysteresis curves in 10.

The hysteresis curves measured at the Fe L_3 edge in **Figure 10** are hardly surprising. Fe is a ferromagnetic material such that the resonant scattering process measures a net moment. In this example the behaviour of the reflectivity nicely follows that of the magnetometry shown in **Figure 8**. It also demonstrates in this case that, the second order terms in Eq. (19) are actually quite negligible. This should not be assumed to be a general case though. More surprising is the ability to be able to measure the magnetic moments in the antiferromagnetic layer i.e. the $IrMn_3$. One possibility is the second order term in Eq. (1). Here the square dependence on the magnetic moment would make it possible to measure the antiferromagnetism. It is exactly this term, in absorption, that is responsible for the magnetic linear dichroism that is often used to measure antiferromagnetic materials. However, in our case this is definitely not the explanation. It would be impossible for the squared dependence to give a hysteresis loop such as those in 10 as the loop would need to have equal reflectivities at both negative and positive saturation. More likely it is due to uncompensated moments near the interface. This could be caused by uncompensated moments domain walls. However, more fundamentally this could just be the effect of measuring the moments near an interface where the moments, even in an antiferromagnetic material do not cancel out.

Measurements of the hysteresis can also be done with linear polarisation. For this we need to work out the equivalent to Eqs. (17) and (18) for linear light. The general result is written out in Eq. (20) for both linear out of the scattering plane σ and linear parallel to the scattering plane π.

$$f = \begin{pmatrix} \sigma_{CR11} & 0 \\ 0 & \pi_{CR22} \end{pmatrix} + i\begin{pmatrix} \sigma_{CI11} & 0 \\ 0 & \pi_{CI22} \end{pmatrix} - i\begin{pmatrix} 0 & \sigma_{M12} \\ \pi_{M21} & \pi_{M22} \end{pmatrix} \tag{20}$$

This will give the general result for σ polarisation

$$\begin{aligned} I &= F_C F_C^* + F_C F_M^* + F_M F_C^* + F_M F_M^* \\ I &= \sigma_{CR11}^2 + \sigma_{CI11}^2 + \pi_{M21}^2 \end{aligned} \tag{21}$$

Here we have simplified the equation since σ polarisation is insensitive to moments out of the scattering plane (we are not taking into account the second order term in 1) and also that all the magnetic scattered x-ray polarisation has flipped π polarisation in agreement with the triple product in the first order term of 1.

For π polarisation Eq. (20) gives us something more complicated.

$$\begin{aligned} I = F_{Total} F_{Total}^* &= F_C F_C^* + F_C F_M^* + F_M F_C^* + F_M F_M^* \\ &= \pi_{CR22}^2 + \pi_{CI22}^2 - 2\pi_{CI22}\pi_{M22} + \pi_{M22}^2 + \sigma_{M12}^2 \end{aligned} \tag{22}$$

Note that Eq. (22) now has a linear and a quadratic term in the π channel as well as a quadratic one in the σ channel. The scattering that rotates into the σ channel is only sensitive to moments in the scattering plane (c.f. Eq. 21). The scattering into the π channel is sensitive to moments perpendicular to the scattering plane. To make this

DOI: http://dx.doi.org/10.5772/intechopen.106831

more readable we can split the equation into two parts: for magnetic moments in the scattering plane

$$I = \pi^2_{CR22} + \pi^2_{CI22} + \sigma^2_{M12} \tag{23}$$

(which looks similar to Eq. (21)) and moments perpendicular to the scattering plane.

$$I = \pi^2_{CR22} + \pi^2_{CI22} - 2\pi_{CI22}\pi_{M22} + \pi^2_{M22} \tag{24}$$

Results from the disordered γ phase sample using both incident linear polarisations σ and π are shown in **Figure 11**. At the top is shown linear horizontal σ and the bottom is shown π incident polarisation. The schematics next to the graphs represent the scattering processes with the moments saturated in the four principal directions. Firstly for σ polarisation, which is only sensitive to moments in the plane, in this case as the moments reduce, so does the reflectivity. It reaches a minimum at the coercive field, and then increases again. Since the dependence on magnetisation is quadratic (see Eq. 21) the reflectivity should be equal for both negative and positive saturation. The small difference is due to locked moments in the ferromagnetic film [37]. Whereas the σ at coercive field shows a minimum, the incident π polarised beam shows a maximum. This is most likely due to the increase of moments and increase in magnetic disorder out of the scattering plane which maximises the expression represented by Eq. (23). Again there is a slight offset possibly caused by some locked moments in the ferromagnetic film.

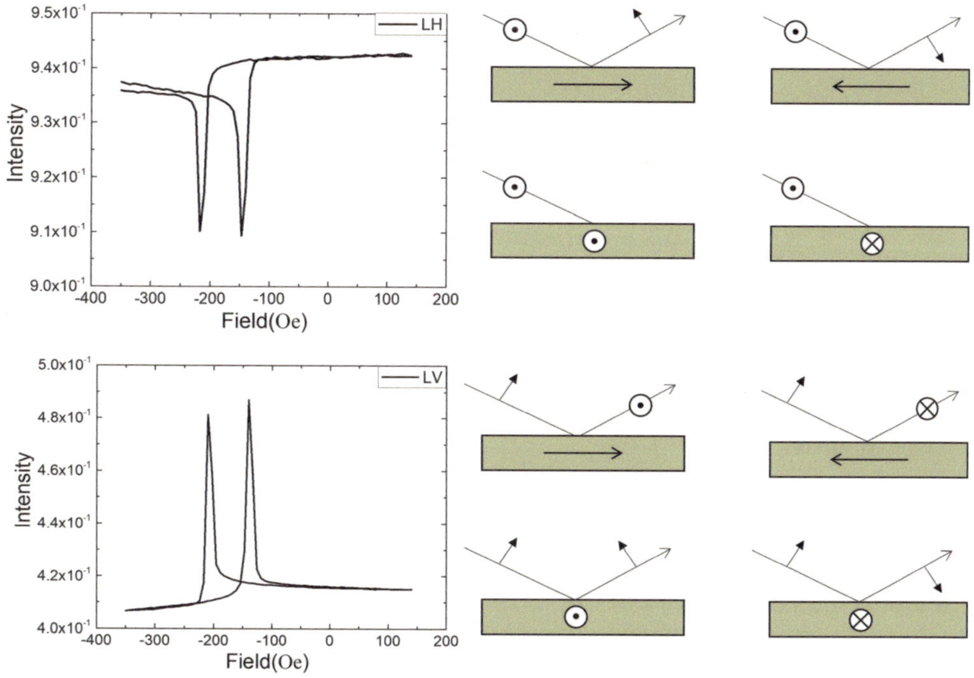

Figure 11.
Magnetic reflectivity hysteresis loops taken at the L_3 edges of Fe on the disordered γ phase $IrMn_3$. At the top is shown linear horizontal and at the bottom linear vertical

6. Note on second order term

In the equation describing the magnetic atomic form factor (Eq. (1)) there are three terms. The last term representing the second order in magnetic moment has been ignored up until now. It is often ignored in most studies due to the assumption that it is small. To measure this in an experiment, particularly with the uncertainty of the coefficients $F^{(0)}$, $F^{(1)}$ and $F^{(2)}$, requires careful exploitation of the polarisation and vector nature of the magnetisation. For example this could be done using polarisation analysis and then measuring magnetic dependence in the σ-σ channel where the polarisation (defined as out of the scattering plane) does not rotate. Since the first order term is zero for this channel any change in the scattering due to manipulation of the moments must be from this second order term. Unfortunately, the inefficient detection in polarisation analysis (less than 1 %) will make this very difficult. A good estimate could also be made without polarisation analysis. Low scattering angles could be used, where all the channels apart from the angle independent σ-σ channel would be weak. Again this could be combined with applied magnetic field dependence so that charge and magnetic scattering can be separated.

Whereas the first order term will provide the first order diffraction peaks from a magnetic lattice the second order term will in addition produce second order satellites. This can easily be demonstrated by inserting a phase factor, for a one dimensional commensurate structure into Eq. (1). In the following we assume that the charge scattering is a real number. Although this is incorrect it simplifies the mathematics and does not influence the main conclusion. If we assume that both the charge and magnetic lattice has a lattice parameter a and insert the corresponding phase factors $\exp iq.na$. Following this, and exactly analogous to Eq. (14), we work out the intensity for a large number of planes N.

$$\begin{aligned} I = N\left(\left((\mathbf{e_f}\cdot\mathbf{e_i})F^{(0)}\right)^2 + \left((\mathbf{e_f}\times\mathbf{e_i})\cdot\mathbf{M}F^{(1)}\right)^2\right)\left(\frac{2\pi}{a}\right)\delta(\mathbf{Q}-\mathbf{G}) \\ +N\left((\mathbf{e_f}\cdot\mathbf{M})(\mathbf{e_i}\cdot\mathbf{M})F^{(2)}\right)^2\left(\frac{2\pi}{a}\right)\delta(2\mathbf{Q}-\mathbf{G}) \end{aligned} \quad (25)$$

Here $\mathbf{G}$ is a reciprocal lattice vector and δ is the Dirac delta function. Eq. (25) shows, in the case of a commensurate structure, that the first two terms of Eq. (1) will give peaks at the the reciprocal lattice vector $\mathbf{Q}$ but the second order term in magnetic moment gives peaks at $2\mathbf{Q}$.

7. Conclusions

This chapter has summarised some of the main techniques in polarised soft X-ray scattering: diffraction, small angle scattering and reflectivity. It has been demonstrated that by tuning to a suitable dipole electric resonance e.g. the $L_{2,3}$ edges of the transition metals or the $M_{4,5}$ edges of the rare earths, which both occur at soft X-ray energies, scattering experiments at these energies are very sensitive to the arrangements of magnetic moments in a material. In diffraction it was demonstrated, for both commensurate and incommensurate magnetic structures, that in addition to the powerful enhancement given by the resonance, the different symmetries of the charge and magnetic lattices allow one to measure purely magnetic signals e.g. in the case of a commensurate antiferromagnet incommensurate helical/spiral spin structures. The relatively new technique of small angle scattering in measuring domains and magnetic skyrmions has been made possible by fabricating ultrathin samples making unprecedented advances in measuring magnetism with a relatively simple experimental setup. The measurement of reflectivity has also been reported on. Here it was shown to be a useful probe for complicated magnetic reversal processes in an exchange biased system, enabling not just the measurement of the ferromagnetic layer but also the antiferromagnet. Measurements with both circular and linear light enabled the probing of the magnetic moments, particularly near the ferromagnetic/antiferromagnetic interface. Circular light is shown to cause interference between charge and magnetic scattering giving a very strong linear magnetic component. Whilst circular light is only sensitive to moments in the scattering plane linearly polarised light is sensitive to moments both parallel and perpendicular to the scattering plane. The sensitive dependence on polarisation means that it is possible to use X-ray scattering as an element specific depth sensitive vector magnetometer.

Acknowledgements

The authors acknowledge J. Herrero-Martin for discussion and the use of figures in section 3.1 on the commensurate antiferromagnetic (AF4) phase in $Mn_{0.85}Co_{0.15}MnWO_4$ The authors also wish to thank P.D. Hatton and M.T. Birch for discussions about the small angle scattering and measuring magnetic skyrmions in section 6.

Conflict of interest

The authors declare no conflict of interest

Author details

Paul Steadman*† and Raymond Fan†
Diamond Light Source Ltd., Didcot, United Kingdom

*Address all correspondence to: paul.steadman@diamond.ac.uk

† These authors contributed equally.

References

[1] Faraday M. Experimental researches in electricity—nineteenth series. Philosophical Transactions of the Royal Society of London. 1846;**136**:1-20

[2] Kerr J. On rotation of the plane of polarization by reflection from the pole of a magnet. The London, Edinburgh, and Dublin Philosophical Magazine and Journal of Science. 1877;**3**(19):321-343

[3] Maxwell JC. A dynamical theory of the electromagnetic field. Philosophical Transactions of the Royal Society of London. 1865;**155**:459-512

[4] Greiner JH, Fan GJ. Longitudinal magneto-optical kerr effect in euo and eus. Applied Physics Letters. 1966;**9**(1):27-29

[5] Krinchik GS, Artem'Ev VA. Magneto-optical properties of ni, co, and fe in the ultraviolet, visible, and infrared parts of the spectrum. Soviet Physics JETP. 1968;**26**(6):1080-1085

[6] Suits JC, Ahn KY. Transverse kerr magneto-optical effect in euo films. Journal of Applied Physics. 1968;**39**(2):570-570

[7] Suits JC, Lee K. Giant magneto-optical kerr effect in euo. Journal of Applied Physics. 1971;**42**(8):3258-3260

[8] Wittekoek S, Bongers PF. Magneto-optical investigation of the band edge of cdcr 2 s 4 and related absorption measurements on cr-doped cdin 2 s 4. IBM Journal of Research and Development. 1970;**14**(3):312-314

[9] De Bergevin F, Brunel M. Observation of magnetic superlattice peaks by x-ray diffraction on an antiferromagnetic nio crystal. Physics Letters A. 1972;**39**(2):141-142

[10] Hannon JP, Trammell GT, Blume M, Gibbs D. X-ray resonance exchange scattering. Physical Review Letters. 1988;**61**(10):1245

[11] Strempfer J, Brückel T, Hupfeld D, Schneider JR, Liss K-D, Tschentscher T. The non-resonant magnetic x-ray scattering cross-section for photon energies up to 500 kev. EPL (Europhysics Letters). 1997;**40**(5):569

[12] Lander GH, Stirling WG. Magnetic x-ray scattering. Physica Scripta. 1992; **T45**:15

[13] Laundy D, Collins SP, Rollason AJ. Magnetic x-ray diffraction from ferromagnetic iron. Journal of Physics: Condensed Matter. 1991;**3**(3):369

[14] McWhan DB, Hastings JB, Kao C-C, Siddons DP. Resonant and nonresonant magnetic scattering. Review of Scientific Instruments. 1992;**63**(1):1404-1408

[15] Blume M, Gibbs D. Polarization dependence of magnetic-x-ray scattering. Physical Review B. 1988;**37**(4):1779-1789

[16] Hill JP, McMorrow DF. Resonant exchange scattering: Polarization dependence and correlation function. Acta Crystallographica Section A: Foundations of Crystallography. 1996;**52**(2):236-244

[17] Herrero-Martín J, Dobrynin AN, Mazzoli C, Steadman P, Bencok P, Fan R, et al. Direct observation of noncollinear order of co and mn moments in multiferroic m n 0.85 c o 0.15 w o 4. Physical Review B. 2015;**91**(22):220403

[18] Hollmann N, Hu Z, Willers T, Bohaty L, Becker P, Tanaka A, et al.

Local symmetry and magnetic anisotropy in multiferroic mnwo4 and antiferromagnetic cowo4 studied by soft x-ray absorption spectroscopy. Physical Review B. 2010;**82**(18):184429

[19] Als-Nielsen J, McMorrow D. Elements of Modern X-ray Physics. Chichester, West Sussex, UK: Wiley; 2001

[20] Chmiel FP, Prabhakaran D, Steadman P, Chen J, Fan R, Johnson RD, et al. Magnetoelectric domains and their switching mechanism in a y-type hexaferrite. Physical Review B. 2019;**100**(10):104411

[21] Alexander J, Johnson RD, Beale TAW, Dhesi SS, Luo X, Cheong SW, et al. Magnetic fan structures in ba0.5sr1.5zn2fe12o22 hexaferrite revealed by resonant soft x-ray diffraction. Physical Review B. 2013;**88**(17):174413

[22] Jamie R, Temple RC, Almeida TP, Lamb R, Peters NA, Campion RP, et al. Asymmetric magnetic relaxation behavior of domains and domain walls observed through the ferh first-order metamagnetic phase transition. Physical Review B. 2020;**102**(14):144304

[23] Fert A, Reyren N, Cros V. Magnetic skyrmions: Advances in physics and potential applications. Nature Reviews Materials. 2017;**2**(7):1-15

[24] Burn DM, Brearton R, Ran KJ, Zhang SL, van der Laan G, Hesjedal T. Periodically modulated skyrmion strings in cu2oseo3. NPJ Quantum Materials. 2021;**6**(1):73

[25] Zhang SL, Bauer A, Berger H, Pfleiderer C, van der Laan G, Hesjedal T. Imaging and manipulation of skyrmion lattice domains in cu2oseo3. Applied Physics Letters. 2016;**109**(19):192406

[26] Zhang S, van der Laan G, Mueller J, Heinen L, Garst M, Bauer A, et al. Reciprocal space tomography of 3d skyrmion lattice order in a chiral magnet. Proceedings of the National Academy of Sciences of the United States of America. 2018;**115**(25):6386-6391

[27] Birch MT, Moody SH, Wilson MN, Crisanti M, Bewley O, Stefancic A, et al. Anisotropy-induced depinning in the zn-substituted skyrmion host cu2oseo3. Physical Review B. 2020;**102**(10):104424

[28] Wilson MN, Crisanti M, Barker C, Stefancic A, White JS, Birch MT, et al. Measuring the formation energy barrier of skyrmions in zinc-substituted cu2oseo3. Physical Review B. 2019;**99**(17):177421

[29] Wilson MN, Birch MT, Štefančič A, Twitchett-Harrison AC, Balakrishnan G, Hicken TJ, et al. Stability and metastability of skyrmions in thin lamellae of Cu_2OSeO_3. Physical Review Research. 2020;**2**:013096

[30] Holm C, Janke W. Critical exponents of the classical 3-dimensional heisenberg-model - a single-cluster monte-carlo study. Physical Review B. 1993;**48**(2):936-950

[31] Zivkovic I, White JS, Ronnow HM, Prsa K, Berger H. Critical scaling in the cubic helimagnet cu2oseo3. Physical Review B. 2014;**89**(6):060401

[32] Vineyard G. Grazing-incidence diffraction and the distorted-wave approximation for the study of surfaces. Physical Review B. 1982;**26**(8):4146-4159

[33] Zak J, Moog ER, Liu C, Bader SD. Fundamental magneto-optics. Journal of Applied Physics. 1990;**68**(8):4203-4207

[34] Zak J, Moog ER, Liu C, Bader SD. Magneto-optics of multilayers with arbitrary magnetization directions. Physical Review B. 1991;**43**(8):6423-6429

[35] Zak J, Moog ER, Liu C, Bader SD. Universal approach to magneto-optics. Journal of Magnetism and Magnetic Materials. 1990;**89**(1-2):107-123

[36] William H, Bean CP. New magnetic anisotropy. Physical Review. 1956;**102**(5):1413

[37] Fan R, Aboljadayel ROM, Dobrynin A, Bencok P, Ward RCC, Steadman P. Dependence of exchange bias on structure of antiferromagnet in fe/irmn3. Journal of Magnetism and Magnetic Materials. 2022;**546**:168678